CARDIOLOGY

International Journal of Cardiovascular Medicine, Surgery and Pathology

Founded 1937 as 'Cardiologia' by Bruno Kisch and Wilhelm Löffler, continued by R. Hegglin (1962–1969), I. Mahaim (1962–1965), P. Moret (1966–1969), P.M. Galletti (1966–1979) and Jan J. Kellermann (1980–1990)

KARGER

Printed in Switzerland
on acid-free paper by
Thür AG Offsetdruck, Pratteln

Appears monthly:
2 volumes per year (12 issues)

Submission: 'Cardiology' publishes original reports in research and observations pertaining to cardiovascular function in health and disease. Only original papers written in English are considered and should be sent to:

Joseph S. Alpert, MD
2601 N. Campbell Avenue
Suite 202A
Tucson, AZ 85719 (USA)
Tel. (602) 326-8334
Fax (602) 326-6188

Manuscripts should be submitted in duplicate (with two sets of illustrations of which one is an original), typewritten double-spaced on one side of the paper, with a wide margin.

Conditions: All manuscripts are subject to editorial review. Manuscripts are received with the explicit understanding that they are not under simultaneous consideration by any other publication. Submission of an article for publication implies the transfer of the copyright from the author to the publisher upon acceptance. Accepted papers become the permanent property of 'Cardiology' and may not be reproduced by any means, in whole or in part, without the written consent of the publisher. It is the author's responsibility to obtain permission to reproduce illustrations, tables, etc. from other publications.

Arrangement

Title page: The first page of each paper should indicate the title (main title underlined), the authors' names, and the institute where the work was conducted. A short title for use as running head is also required.

Key words: For indexing purposes, a list of 3–10 key words in English is essential.

Abstract: Each paper needs an abstract of up to 10 lines.

Small type: Paragraphs which can or must be set in smaller type (case histories, test methods, etc.) should be indicated with a 'p' (petit) in the margin on the left-hand side.

Footnotes: Avoid footnotes. When essential, they are numbered consecutively and typed at the foot of the appropriate page.

Tables and illustrations: Tables and illustrations (both numbered in Arabic numerals) should be prepared on separate sheets. Tables require a heading and figures a legend, also prepared on a separate sheet. For the reproduction of illustrations, only good drawings and original photographs can be accepted; negatives or photocopies cannot be used. When possible, group several illustrations on one block for reproduction (max. size 144 × 196 mm) or provide crop marks. On the back of each illustration, indicate its number, the author's name, and 'top'. Color illustrations are reproduced at the author's expense.

References: In the text identify references by Arabic numerals [in square brackets]. Material submitted for publication but not yet accepted should be noted as 'unpublished data' and not be included in the reference list. The list of references should include only those publications which are cited in the text. Do not alphabetize; number references in the order in which they are first mentioned in the text. The surnames of the authors followed by initials should be given. There should be no punctuation other than a comma to separate the authors. Cite all authors, 'et al' is not sufficient. Abbreviate journal names according to the Index Medicus system. (Also see International Committee of Medical Journal Editors: Uniform requirements for manuscripts submitted to biomedical journals. Br Med J 1988;296:401–405.)

Examples

(a) Papers published in periodicals: Kauffman HF, van der Heide S, Beaumont F, Blok H, de Vries K: Class-specific antibody determination against *Aspergillus fumigatus* by means of the enzyme-linked immunosorbent assay. III. Comparative study: IgG, IgA, IgM ELISA titers, precipitating antibodies and IgE binding after fractionation of the antigen. Int Arch Allergy Appl Immunol 1986;80:300–306.

(b) Monographs: Matthews DE, Farewell VT: Using and Understanding Medical Statistics. Basel, Karger, 1985.

(c) Edited books: Hardy WD Jr, Essex M: FeLV-induced feline acquired immune deficiency syndrome: A model for human AIDS; in Klein E (ed): Acquired Immunodeficiency Syndrome. Prog Allergy. Basel, Karger, 1986, vol 37, pp 353–376.

Full address: The exact postal address complete with postal code of the senior author must be given. If correspondence is handled by someone else, indicate this accordingly.

Further details: Authors may consult our leaflet 'Rules for the Preparation of Manuscripts', available on request from the Publisher, S. Karger AG, P.O. Box, CH-4009 Basel (Switzerland).

Page charges: There is no page charge for papers of 8 or fewer printed pages (including tables, illustrations and references). Each additional complete or partial page is charged to the author at SFr. 225.– / US $ 150.00. The allotted size of a paper is equal to approx. 16 manuscript pages (including tables, illustrations and references).

Galley proofs: Unless indicated otherwise, galley proofs are sent to the first-named author and should be returned with the least possible delay. Alterations made in galley proofs, other than the correction of printer's errors, are charged to the author. No page proofs are supplied.

Reprints: Order forms and a price list are sent with the galley proofs. Orders submitted after the issue is printed are subject to considerably higher prices. Allow five weeks from date of publication for delivery of reprints.

KARGER

CARDIOLOGY

Publication data: 'Cardiology' is published 12 times annually. Volumes 82, 83 with 6 issues per volume appear in 1993.

Subscription rates: Subscriptions run for a full calendar year. Prices are given per volume.

Personal subscription:

Subscription	*+ postage and handling*	
SFr. 71.50	+ SFr. 3.90	Europe
SFr. 71.50	+ SFr. 5.90	Overseas
US$ 47.50	+ US$ 3.90	

(Must be in the name of, billed to, and paid by an individual. Order must be marked 'personal subscription'.)

Institutional subscription:

Subscription	*+ postage and handling*	
SFr. 439.–	+ SFr. 24.–	Europe
SFr. 439.–	+ SFr. 36.–	Overseas
US$ 293.00	+ US$ 24.00	

Airmail postage: SFr. 36.–, US$ 30.00. (Extra per volume.)

Microform edition: (Microfilms 16 and 35 mm or microfiches) 50% discount on the regular subscription rate for subscribers of the printed edition.

Advertising: Correspondence and rate requests should be addressed to the Publisher.

Subscription orders: Orders can be placed at agencies, bookstores, directly with the Publisher S. Karger AG, P.O. Box, CH – 4009 Basel (Switzerland) or with any of the following distributors:

Deutschland: S. Karger GmbH
Lörracher Strasse 16A, Postfach
D-W–7800 Freiburg

France: Librairie Luginbühl
36, bd de Latour-Maubourg
F–75007 Paris

Great Britain, Ireland: S. Karger Publishers
11 Ravenscroft Road
Beckenham, Kent BR3 4TP

USA: S. Karger Publishers, Inc.
26 West Avon Road, P.O. Box 529
Farmington, CT 06085

India, Bangladesh, Sri Lanka:
S. Karger India Subscription Agency
B – 5/132 Safdarjung Enclave
New Delhi-110029

Japan: Katakura Libri, Inc.
36-9 Hongo 3-chome, Bunkyo-ku
Tokyo 113

Change of address: Both old and new address should be sent to the subscription source.

Single issues and back volumes: Information can be obtained through the Publisher.

Bibliographic indices: This journal is regularly listed in bibliographic services, including *Current Contents®*.

Book Reviews: Books are accepted for review by special agreement.

KARGER

Contribution of Long-Term Follow-Up to the Prediction of Coronary Heart Disease

Guest Editors
M. Kornitzer, Bruxelles
R. Goldberg, Worcester, Mass.

24 figures and 67 tables, 1993

KARGER

S. Karger
Medical and Scientific Publishers
Basel · Freiburg · Paris · London
New York · New Delhi · Bangkok
Singapore · Tokyo · Sydney

Drug Dosage
The authors and the publisher have exerted every effort to ensure that drug selection and dosage set forth in this text are in accord with current recommendations and practice at the time of publication. However, in view of ongoing research, changes in government regulations, and the constant flow of information relating to drug therapy and drug reactions, the reader is urged to check the package insert for each drug for any change in indications and dosage and for added warnings and precautions. This is particularly important when the recommended agent is a new and/or infrequently employed drug.

Printed in Switzerland on acid-free paper by
Thür AG Offsetdruck, Pratteln
ISBN 3–8055–5790–6

Contents

Cardiology 1993;82:85–88

M. Kornitzer
R. Goldberg

Introduction

In 1948–1950, the first prospective study investigating the determinants of coronary heart disease was initiated in the town in Framingham, Massachusetts, in the United States of America [1]. Over the following decade, numerous work site- and population-based prospective cardiovascular studies were begun in the US, Europe and Israel, including the multinational Seven Countries Study.

After 5–10 years of follow-up, three major modifiable coronary risk factors emerged in most of these studies as independent predictors of coronary heart disease (CHD): serum cholesterol, blood pressure and cigarette smoking. Several of these studies initiated in middle-aged subjects have now reached follow-up periods of between 20 and 30 years.

It seemed to us appropriate and timely to attempt to present in a single issue results of several of these landmark studies in terms of the incidence rates of, and mortality from, CHD and all-cause mortality, in relation to these and other predictive factors.

These long-term follow-up studies provided an opportunity to address a number of questions including the following.

(1) Are the predictors for comparatively short-term increased risk for CHD still predictive in an aging population? In other words, are middle-aged men and women at increased risk for CHD during essentially the remainder of their life from measures carried out approximately 20–30 years earlier? A positive response to this question should have a great impact on both public health policy and clinical practice. A related question which could be asked is: Are the principal coronary risk factors also long-term predictors of all-cause mortality (are these 'nonspecific' risk factors)? Again, an affirmative response to this question should have far-reaching public health implications.

(2) Another question concerns the predictive power of the major coronary risk factors measured in subjects aged 60 and over.

(3) Finally, a third question concerns the biological variability of most of these measures leading to a regression dilution bias [2] and consequently weakening any observed associations between exposure and disease.

Most of the papers appearing in this issue tackled the first question. Table 1 provides an overview of the results of these studies. Some papers concern more than one cohort: the Evans County Heart Study reports data on black and white males and females. The article by Stamler et al. describes the findings from three different Chicago studies ranging

Prof. M. Kornitzer
Ecole de Santé Publique
U.L.B.-Campus Erasme CP 597
Route de Lennik 808
B-1070 Bruxelles (Belgium)

0008-6312/93/
0823-0085$2.75/0

Table 1. Long-term independent predictors of CHD incidence and/or mortality

Authors and studies	Patients	FU years	Serum cholesterol	Blood pressure	Cigarette smoking	Diabetes	Nutrition
De Vries et al[a] (Zutphen Study)	677	20	+	+	NA	NR	NR
Goldbourt et al. (Israel Ischemic Heart Disease Study)	10,059	23	+	+	+	+	saturated fats P/S –
Hames et al. (Evans County Heart Study)	3,102	20					
White males			+	+	+	NR	NR
Black males			+	+	+	NR	NR
White females			+	+	NR	NR	NR
Black females			+	+	NR	NR	NR
Kannel and Larson (Framingham Study)		long term[b]					
Males			+	+	+	+	NR
Females			+	+	+	+	NR
Kornitzer et al. (Belgian Bank Study)	1,227	25	+	+	+	NR	NR
Menotti et al.[a] (Italian Rural Cohorts)	1,712	25	+	+	+	NR	NR
Pekkanen et al. (Finnish Cohorts)	1,619	30	+	+	+	NR	NR
Stamler et al.							
Western Electric	1,903	24	+	+	+	NR	dietary chol. + saturated fats NA
Peoples Gas	1,405	30	NA	+	+	NR	NR
Chicago Heart Association Detection Project							
Males (40–59)	8,515	15	+	+	+	+	NR
Females (40–59)	7,082	15	+	+	+	+	NR
Males (60–74)	1,490	15	+	+	+	+	NR
Females (60–74)	1,243	15	+	+	+	+	NR
Welin et al.[a] (Göteborg Study)	845	25	+	+	+	NR	NR

[a] Incidence of hard events (non fatal, fatal myocardial infarction, sudden death).

+ = Significant predictor; – = significant inverse association; NA = no association; NR = not relevant (not measured); FU = follow-up.

[b] Long term (no more precision in the paper).

from 15 to 30 years of follow-up. The Framingham Study includes over 5,000 males and females free from CHD at the time of the baseline examination.

As can be seen, all studies have shown with impressive consistency the independent association of the three major coronary risk factors, namely serum cholesterol, blood pressure and cigarette smoking, with CHD. The relation is observed for males and females, blacks and whites, different religious denominations, and in US and European cohorts. There are very rare exceptions to these findings: smoking is not associated with the 20-year incidence of CHD in the Zutphen Study (De Vries et al.). Blood pressure is only associated with mortality from CHD during the first 20 years of follow-up, and serum cholesterol is not associated with 30-year CHD mortality in the People's Gas cohort (Stamler et al.).

In those studies where diabetes was determined at the baseline examination, an independent association with CHD has been observed (Goldbourt et al., Kannel and Larson, Stamler et al. Chicago Heart Association Detection Project).

Two studies report the independent prediction of CHD for several nutritional variables: In the Israeli Ischemic Heart Disease Study, Goldbourt et al. report that saturated fat intake, expressed as a percentage of caloric intake, as well as the P/S ratio were related to 23-year CHD mortality; in the Western Electric Company Study, Stamler et al. report a significant association for dietary cholesterol with 24-year deaths from CHD. The relationship between a secondary risk factor for CHD, obesity and CHD mortality is less consistent: overweight is an independent predictor for long-term death from CHD in some US studies (Framingham, Chicago Western Electric) and Finnish cohorts (Pekkanen et al.), but not in the Italian cohorts of the Seven Countries Study, (Menotti et al.), the Belgian Bank Study (Kornitzer et al.), and the Israeli Ischemic Heart Disease Study (Goldbourt et al.). Goldbourt et al. highlight the importance of HDL cholesterol as a long-term independent predictor of CHD mortality, confirming earlier findings from the Framingham study. In these studies, high blood pressure and cigarette smoking are consistently shown to be independent long-term predictors for all-cause mortality while the results for serum cholesterol are less consistent.

As for the second question posed at the outset of this editorial, two papers in this issue report an association of the major established coronary risk factors with morbid or fatal events in older subjects. In the Chicago Heart Association Detection Project, serum cholesterol, blood pressure and smoking are associated with CHD death in males and females aged 60–74 at baseline. The Framingham Study also reports an association of serum cholesterol and blood pressure with CHD incidence in males and females aged 65–90.

The third question raised in this editorial has been debated for a long time. Is there any necessity to either obtain multiple baseline measures or serial measurements over time for specific biological variables in order to reduce the recently postulated regression dilution bias [3, 4]? By meta-analysis, McMahon et al. [2] have shown that the regression dilution bias indeed decreases the real association observed between coronary risk factors and CHD events.

In this issue, De Vries et al. have confirmed that the average of repeated measures over time increases the statistical predictive power of both serum cholesterol and systolic blood pressure. On the other hand, Menotti et al. show in their paper that, despite the likelihood of classification bias with a single measure of serum cholesterol and blood pressure at baseline, an increase or decrease in these variables over time is independently related to 20-year CHD incidence above and beyond the baseline value of these variables.

In summary, the principal findings in this issue are that serum cholesterol, blood pressure and cigarette smoking are long-term independent predictors for CHD incidence or mortality; blood pressure and cigarette smoking are consistently observed to be long-term predictors of all-cause mortality. These observations run across culture, religion affiliations, ethnic orgins, age and sex. Those who advocate a structured public health policy at the community level in order to promote the primary prevention of CHD and reduce the population burden from coronary atherosclerosis should be comforted by these reports. Clinicians should also find considerable food for thought in these collective findings. The detection of individuals at high risk for CHD in early adulthood by opportunistic screening or 'case finding' should be greatly encouraged with possibilities not only for the modification of adverse life-style habits related to CHD but also to total mortality.

References

1 Dawber T, Meadors G, Moore M, Moore F: Epidemiological approaches to heart disease: The Framingham Study. Am J Public Health 1951;41:279–286.
2 MacMahon S, Peto R, Cutler J, Collins R, Sorlie P, Neaton J, Abbot R, Godwin J, Dyer A, Stamler J: Blood pressure, stroke and coronary heart disease. I. Prolonged differences in blood pressure: Prospective observational studies corrected for the regression dilution bias. Lancet 1990; 335:765–774.
3 Kahn H, Dawber T: The development of coronary heart disease in relation to sequential biennial measures of cholesterol in the Framingham Study. J Chronic Dis 1966; 19:611–620.
4 Truett J, Sorlie P: Changes in successive measurements and the development of disease: The Framingham Study. J Chronic Dis 1971; 24:349–361.

Cardiology 1993;82:89–99

Christina L. de Vries
Edith J.M. Feskens
Cor de Lezenne Coulander
Daan Kromhout

Department of Epidemiology,
National Institute of Public Health
and Environmental Protection,
Bilthoven, The Netherlands

Repeated Measurements of Serum Cholesterol and Blood Pressure in Relation to Long-Term Incidence of Myocardial Infarction

The Zutphen Study

Key Words
Repeated measurements
Cholesterol
Systolic blood pressure
Follow-up
Cohort
Myocardial infarction incidence

Abstract
Repeated risk factor measurements reduce the amount of random error due to intraindividual variation. Annual measurements of serum cholesterol and systolic blood pressure were done in a cohort of 878 men, aged 40–59 years, between 1960 and 1970. This study examined for both risk factors the association with the incidence of myocardial infarction (MI) during follow-up till 1985. Mean values from repeated measurements yielded stronger statistical associations with MI incidence than single observations. However, when measurements were extended over too many years in this middle-aged cohort, a bias occurred due to selective loss of high-risk individuals as early cases. This was especially the case for serum cholesterol. Taking account of trends in regression analysis did not show any advantage over the use of mean values.

'An epidemiologic study is properly viewed as an exercise in measurement, with accuracy as the goal' [1]. *K.J. Rothman,* 1986

Introduction

In cohort studies, usually considerable effort is made to measure the outcome variable, e.g. myocardial infarction (MI), as accurate as possible. Of equal importance is the accurate measurement of independent variables of risk factors. Standardization procedures reduce systematic observer and laboratory measurement errors. Random error will still occur, due to inter- and intraindividual variation, and between-batch variation measurement errors [2]. In various prospective cohort studies, a single baseline value for risk factor

Edith J.M. Feskens, PhD
Department of Epidemiology
National Institute of Public Health and
Environmental Protection, PO Box 1
NL-3720 BA Bilthoven (The Netherlands)

0008-6312/93/
0823-0089$2.75/0

assessment is used [3]. Also, previous reports on the Zutphen Study were concerned with the predictive value of a single screening test in relation to the occurrence of MI [4]. This can be valid and of importance, when the point of the analysis is to test the predictiveness of a measurement, which will be a function of both the underlying biological relationship as well as the validity of the measurement [5]. However, if the focus is to assess the magnitude of the biological relationship between serum cholesterol, blood pressure and long-term incidence of MI, then consideration should be given to correcting for random error.

With follow-up periods extended over several decades, risk factor levels may change due to secular trends and life-style changes related to period and/or age and due to random changes. In other words, with increasing time, more within-person variance will be present around the level of a risk factor. The importance of intraindividual variation for cholesterol was quantified in a Swedish prospective cohort study: the correlation between cholesterol assessments at 6-week intervals was found to be 0.75 and at 2-year intervals 0.66 [6]. For repeated measurements amongst 14,000 subjects in the UK, the within-person coefficient of variation (CV) was 7.4% for measurements 1 year apart [7]. For blood pressure the biological intraindividual variation is even greater, and the potential for misclassification of subjects results in bias of measures of association towards the null hypothesis and leads to an underestimation of the true risk ratios [1]. This may become particularly a problem in etiologic research in epidemiology, where risk ratios are small, and in cohort studies with long follow-up periods.

This paper deals with the value of repeated risk factor measurements above the use of a single baseline measurement. This is evaluated in the Zutphen Study. An important feature of this prospective cohort study is the availability of data on annual blood pressure and serum cholesterol measurements of the majority of the study subjects during the first 10 years of follow-up. These data provide the opportunity to analyze the intraindividual variability of serum cholesterol and blood pressure measurements for this cohort. The following hypotheses were tested. (1) Does the use of annually repeated measurements of serum cholesterol and blood pressure improve the predictive value of these risk factors in relation to MI incidence? (2) Is a predicted value, based on the trend of several measurements in the preceding years (linear regression analysis), closer associated with MI incidence than the simple average of the same measurements? (3) Do four to eleven measurements over a 10-year period give better results than four to five measurements in 5 years?

Subjects and Methods

Since 1960, a longitudinal investigation of risk factors for chronic diseases was carried out among middle-aged men from the town of Zutphen in The Netherlands. The Zutphen Study is the Dutch contribution to the Seven Countries Study [3, 9]. In 1960, a random sample of 1,088 men born between 1900 and 1919 was drawn from a total of 2,450. A cohort of 878 men from 40 to 59 years was eventually medically examined.

Serum cholesterol determinations were carried out according to the method of Abell et al. [10] in the period 1960–1965 and the method of Anderson and Keys [11] between 1966 and 1970. The latter method provides values equivalent to the method of Abell et al. [10]. The average serum cholesterol over the periods (a) 1960–1965 and (b) 1960–1970 was calculated as the sum of the cholesterol readings divided by the number of measurements, provided that the subject had a total of at least four readings during this period. For the period 1960–1970, at least one of these blood samples should have been taken in the last 5 years prior to or in 1970. In order to include trends in serum cholesterol levels, the predicted serum cholesterol level was obtained by linear regression of the serum cholesterol readings from 1960 to 1965 or 1970 on time (years).

Blood pressure was measured annually between 1960 and 1970. Measurements were carried out with an ordinary mercury sphygmomanometer at the end of the physical examination on the right arm with the men in supine position, according to the protocol of the Seven Countries Study [3]. The average systolic blood pressure (SBP) was calculated analogous to cholesterol as described above, also provided that the subject had a total of at least four readings during this period, and for the period 1960–1970, at least one of the readings was from the last 5 years prior to or in 1970.

Exposure to cigarette smoking was summarized as the product of the average number of cigarette smoked per day and the number of smoking years (cigarette years).

During follow-up, annual medical examinations of the men took place between 1960 and 1973, and after that only once in 1977/1978. In 1980 and 1982, a questionnaire was filled in by the men; in 1985 a Dutch translation of the questionnaire of Rose and Blackburn [12] was used. Information was obtained from the general practitioners of the subjects and from hospital discharge forms. No one was lost to follow-up. All morbidity and mortality data collected in the period 1960–1985 were coded by one physician in a standardized way in 1986 (according to the 8th revision of the International Classification of Diseases). The diagnosis MI was based on three criteria: a specific medical history, i.e. severe chest pain lasting for more than 200 min and not disappearing in rest, characteristic ECG changes (major or lesser Q waves and major T wave findings) and specific enzyme elevations. MI was assumed to be present when two of the three criteria was fulfilled, or otherwise when MI was denoted as primary cause of death on the death certificate. Data on MI incidence, being either the first nonfatal or fatal event, are presented here.

From the 755 men who were alive in 1970, 130 were excluded either because of prevalence of MI before 1970, death in 1970 or no physical examination or blood sample obtained in 1970. From the 625 remaining men, the study group was formed by 555 subjects (89%) of whom four or more measurements were available for both serum cholesterol and blood pressure. Eleven readings for all-risk factors were available for 439 men (79%) amongst the study group. The same criteria resulted in 677 eligible men in the study group of 1965 (77% of the original cohort).

The results of at least four measurements between 1960 and 1965 and between 1960 and 1970 on the incidence of MI during the years of follow-up till 1985 were studied in uni- and multivariate analyses. Statistical analyses were performed using the SPSS/PC+ and BMDP package programs [13, 14]. MI risk was assessed by survival analysis, using the Cox proportional hazards model [15]. The risk factors serum cholesterol and SBP were entered as categorical variables (quartiles) in uni- and multivariate analysis models. Age was used as a continuous variable.

The intraindividual variance was estimated by calculating the residual variance around the individual regression line of the different measurements of variables on time [16]. For comparative purposes, the estimated intraindividual variances were also expressed as CV [17], by dividing the variances by individual mean levels.

Results

Among the cohort of 677 men (aged 45–64 years in 1965), there were 134 new cases diagnosed with MI (46 fatal and 88 nonfatal) during 20 years of follow-up between 1965 and 1985. The overall incidence rate was 12.5/1,000 person years.

Amongst the cohort of 555 men (aged 50–69 years in 1970), there were 90 new cases diagnosed with MI (29 fatal and 61 nonfatal) during 15 years of follow-up between 1970 and 1985. The overall incidence rate was 12.9/1,000 person years.

The measured values in 1960, 1965 and 1970, the mean values over the period 1960–1965 and 1960–1970, and the predicted values for 1965 and 1970 (estimated by applying the linear regression technique) for serum cholesterol and SBP are presented in table 1. There was no statistically significant age difference between the group that developed an MI, and the group that did not. The risk factor averages for the group of men who developed an MI during follow-up were slightly higher at all times and with each method used than for the group of men without MI during follow-up. The p values for single observations in

Table 1. Observed, mean and predicted values of serum cholesterol and SBP in 1960, 1965 and 1970 (± SD) in relation to MI incidence up to 1985: the Zutphen Study

Risk factor	Without MI	With MI	p value
Cholesterol, mmol/l			
Observed in 1960	6.02 ± 1.18	6.32 ± 1.10	0.002
Observed in 1965	5.98 ± 1.03	6.18 ± 0.92	0.04
Observed in 1970	6.11 ± 1.12	6.31 ± 0.94	0.2
Mean 1960–1965	6.00 ± 0.99	6.26 ± 0.91	0.005
Mean 1960–1970	6.07 ± 0.96	6.17 ± 0.73	0.3
Predicted for 1965	5.91 ± 1.02	6.14 ± 0.90	0.01
Predicted for 1970	6.17 ± 1.05	6.31 ± 0.79	0.1
SBP, mm Hg			
Observed in 1960	142 ± 19	149 ± 21	0.0001
Observed in 1965	141 ± 17	146 ± 19	0.02
Observed in 1970	145 ± 20	149 ± 21	0.08
Mean 1960–1965	138 ± 13	143 ± 15	0.002
Mean 1960–1970	141 ± 14	145 ± 15	0.02
Predicted for 1965	138 ± 16	141 ± 16	0.02
Predicted for 1970	147 ± 19	152 ± 18	0.01
Age, years	54.6 ± 5.5	55.0 ± 5.2	0.4
Subjects			
1960	637	181	
1965	543	134	
1970	465	90	

Total number of subjects in 1960, 1965 and 1970 were 818, 677 and 555, respectively. The t test was applied.

1960, 1965 and 1970 show a trend being smallest in 1960 and larger in 1970. The mean intraindividual CV for serum cholesterol in this cohort was 8.1% (SD 2.4%), the mean CV for SBP was 6.9% (SD 2.0%).

For serum cholesterol in 1970, differences were nonsignificant, regardless of the method used (p values >0.1). The differences in means for repeated serum cholesterol measurements 1960–1965 showed a considerably smaller p value than for single measured serum cholesterol in 1965 (0.005 vs. 0.04). For SBP the observed difference in means in 1970 was not statistically significant, however the mean 1960–1970 values and the predicted value did become significant at the 5% level.

Table 2. Cutoff points for quartiles of serum cholesterol and SBP according to method and year of assessment (for 1965 and 1970 also the means ± SD for each quartile are shown.): the Zutphen Study

Risk factor	P25	P75	Mean lowest quartile	Mean middle quartiles	Mean highest quartile
Cholesterol, mmol/l					
Measured 1960	5.30	6.71			
Measured 1965	5.33	6.68	4.80	6.00	7.35
Mean 1960–1965	5.38	6.60	4.88	5.99	7.32
Predicted 1965	5.24	6.56	4.73	5.90	7.26
Measured 1970	5.39	6.78	4.84	6.13	7.54
Mean 1960–1970	5.44	6.66	4.96	6.04	7.30
Predicted 1970	5.48	6.84	4.96	6.15	7.52
SBP, mm Hg					
Measured 1960	130.0	155.0			
Measured 1965	130.0	150.0	123	142	166
Mean 1960–1965	129.3	146.3	124	137	158
Predicted 1965	127.4	146.4	120	136	160
Measured 1970	130.0	156.0	124	144	175
Mean 1960–1970	132.1	148.9	126	140	161
Predicted 1970	134.2	158.3	127	146	173

In 1965, means for SBP had a smaller p value than for the observed values as well.

Intervals between mean values of quartiles were smaller than for measured values reflecting 'regression to the mean'. The standard deviations were also considerably smaller for mean values than for predicted and single measured values, reflecting the substantial effect of reducing the intraindividual variation in repeated measurements on the total variance (the sum of both intra- and interindividual variability).

The subjects were classified into quartiles for each risk factor and method. Thus quartile cutoff values (P25 and P75) varied according to the method used. This also applied to the mean for lowest and highest quartiles, as shown in table 2.

The relative risks (RR), as found in multivariate analysis, are shown for 1965 in table 3. For serum cholesterol, RR for the highest versus lowest quartile was 2.5, both for single measured values and mean values. Only the p value was slightly smaller for RR according to mean values. The predicted value for cholesterol showed a poorer association with MI incidence.

SBP was statistically significantly associated with MI irrespective of the method used, however the mean values over a 5-year period

Table 3. RR for MI incidence during 20 years of follow-up in multivariate survival analysis: the Zutphen Study

Risk factor	Hazard ratio	95% confidence intervals	p value
Age	1.04	1.00–1.08	0.06
Cholesterol			
Observed in 1965			
<P25	1		
P25–P75	2.53	1.51–4.22	
>P75	2.49	1.42–4.37	0.003
Mean 1960–1965			
<P25	1		
P25–P75	2.17	1.29–3.64	
>P75	2.52	1.45–4.38	0.002
Estimated for 1965			
<P25	1		
P25–P75	1.97	1.21–3.20	
>P75	1.81	1.05–3.10	0.05
SBP			
Observed in 1965			
<P25	1		
P25–P75	1.86	1.18–2.93	
>P75	2.10	1.27–3.48	0.005
Mean 1960–1965			
<P25	1		
P25–P75	1.92	1.13–3.24	
>P75	3.18	1.84–5.52	0.0001
Estimated for 1965			
<P25	1		
P25–P75	1.88	1.13–3.13	
>P75	2.75	1.59–4.77	0.0003
Cigarette years till 1965			
<P25	1		
P25–P75	1.34	0.86–2.08	
>P75	1.29	0.79–2.10	0.4

Separate models by different methods of measurement of serum cholesterol, SBP and cigarette years are shown. Hazard ratio, calculated from Cox proportional hazards model; p value: score statistics for linear trend.

showed the strongest association. RR for the highest versus the lowest quartile was 3.2 for mean SBP versus 2.1 for measured SBP values (in 1965). Similar to serum cholesterol, the predicted value was not so strongly significantly associated with MI incidence.

Smoking showed only a small risk ratio (RR = 1.3), being statistically not significant in each of the models.

The RR in 1970 for serum cholesterol and SBP were overall lower than in 1965. Table 4 shows the differential loss of subjects in the highest quartiles during the period 1960–1970. Loss occurred both due to MI incidence during the period 1960–1970 and death from all causes. After 1970, the difference in MI incidence (in percentage) between the quartiles was smaller than before 1970.

Discussion

Initially, information on repeated, annual measurements over a 10-year period (1960–1970) was used to investigate the effect of intraindividual variability on misclassification of serum cholesterol and SBP in a cohort study. However, the results in tables 1 and 3 were poor, particularly for serum cholesterol, and did not indicate a clear risk gradient. Means per quartile of ascending serum cholesterol values for a 10-year period showed that risks for MI were highest for the P50–P75 quartile of serum cholesterol. After reviewing the data, an important shortcoming of our approach was found which could largely explain these findings. One should bear in mind that all prevalent and incident cases between 1960 and 1970 were excluded. The early MI cases (1960–1970) amongst middle-aged men (age ranging from 40 to 69 years) came largely from the highest quartile risk groups: 15.9 against 3.8% in the lowest quartile, leaving others with less extreme risk profiles for the

Table 4. Serum cholesterol and SBP quartiles in 1960 and MI incidence and death from all causes of 829 men during 25 years of follow-up (1960–1985): the Zutphen Study

Follow-up period	1960–1970		1970–1985	
	MI	all causes	MI	all causes
SBP				
<P25	4.0	8.2	11.6	32.6
P25–P75	5.4	8.5	6.6	36.7
>P75	9.7	21.5	16.9	41.0
Cholesterol				
<P25	2.5	10.0	9.4	33.3
P25–P75	6.9	13.0	14.0	38.2
>P75	7.1	10.6	13.0	35.8

Values are given in percent.

follow-up period starting in 1970. Therefore, it was possible to find the highest association between the P50 and P75 quartile of serum cholesterol and late-onset MI. In figure 1 it is apparent in the survival curves how the high-risk group has a surplus case rate in the first 10 years, after which the three highest quartiles have similar risks, but distinct from the baseline group. The P25 cutoff point for mean serum cholesterol was 5.38 mmol/l. This confirms the usefulness of the NCEP guidelines which recommend an upper limit for a desirable average of three cholesterol measurements at 5.4 mmol/l [18].

The results presented for the period 1960–1965 were in accordance with the results reported by Reed and Benfante [19]. Similar to the Zutphen cohort, birth in a cohort of men in Hawaii with a follow-up period of 20 years, serum cholesterol measured at middle age was found to predict both early- and late-onset coronary heart disease. They also found that the mean of three or four measurements of serum cholesterol provided a slightly better estimate of the relative risk for long-term MI incidence than a single measurement.

Results according to SBP quartiles were even more pronounced: 24.9% of the highest quartile as assessed in 1960 died before 1970 against 9.9% from the lowest quartile in the same period. After 1970, the death rate in the third quartile is equally high as in the highest quartile (table 4). Again a certain amount of 'selection of the fittest' occurred. Repeated measurements over a 5-year period yielded better results compared to a 10-year period. Apparently, the bias due to selective loss of high-risk individuals in the first 10 years of follow-up was more important than the gain in precision. The mean SBP (1965) was associated closest with MI incidence. SBP was a stronger risk factor than serum cholesterol, and rate ratios increased significantly and according to ascending risk group. In contrast with serum cholesterol, the survival curves according to SBP quartiles ran away from each other under a fairly constant angle throughout the whole follow-up period, as can be seen in figure 2.

The average CV_a for serum cholesterol (8.1%) was comparable to the one (7.4%) found in the UK [7], The Netherlands (6.5%) [20] and a second cohort study in The Nether-

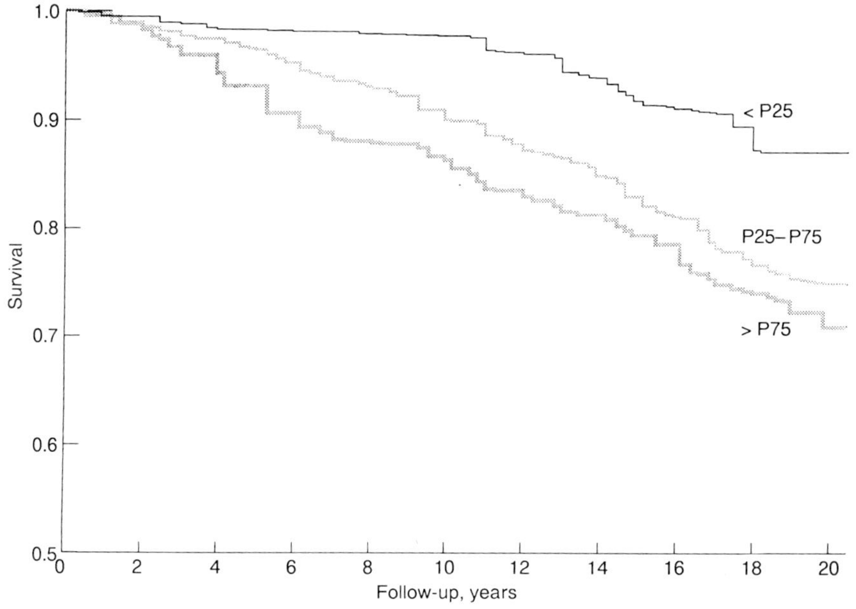

Fig. 1. Incidence of MI during 20 years of follow-up by mean serum cholesterol (1960–1965): the Zutphen Study. Follow-up is started in 1965 (year 0). Mean serum cholesterol is classified according to quartiles: lowest quartile (<P25); middle quartiles (P25–P75), and highest quartile (>P75).

lands (9%) [21]. In this last-mentioned study, conducted amongst elderly men and women in Rotterdam, 6.8% were found for the CV_a for SBP, which is approximately the same as the 6.9% for SBP found in the present study. Such CV can contribute substantially to the attenuated association between risk factors measured on a single occasion and long-term MI incidence as follows from this study and from several other studies [5, 22–26].

To take account of trends in the risk factor over a 5- or 10-year period, by applying linear regression, did not result in a closer or stronger association with MI incidence than taking just the mean value. This was true for both risk factors examined and for both the 5- and 10-year period. This was also reported for serial SBP measurements in relation to the risk for stroke in this same cohort [27]. An explanation could be that prediction intervals are much wider than 95% CV for a single measurement: both the individual error and the error in estimating the new mean need to be taken in account [28]. This is expressed in the larger standard deviations around the predicted valued than around the mean values.

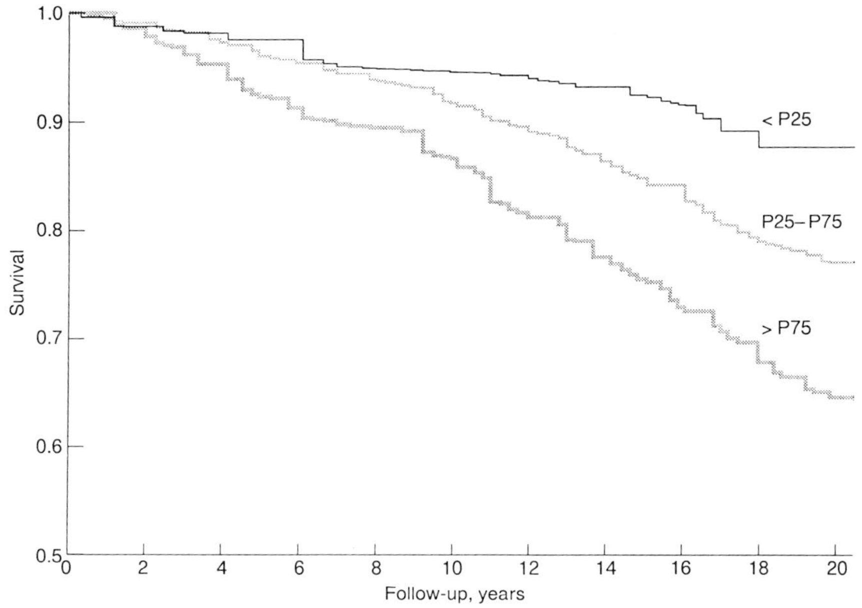

Fig. 2. Incidence of MI during 20 years of follow-up by mean SBP (1960–1965): the Zutphen Study. Follow-up is started in 1965 (year 0). Mean SBP is classified according to quartiles: lowest quartile (<P25); middle quartiles (P25–P75), and highest quartile (>P75).

Comparison of the uni- and multivariate analyses showed that cigarette smoking did not confound the associations between serum cholesterol, SBP and incidence of MI. A problem with the variables for smoking was that, in this cohort of men, 92.6% were smokers (1960). There was no homogeneous baseline group of only nonsmokers. Hence no significant association between smoking and MI was noticed. In the present study, this large smoking prevalence could also be viewed as an advantage, as the effect of cholesterol and blood pressure could be studied better when virtually all subjects were smokers.

Conclusions

In this cohort study, a stronger association was obtained between the mean of repeated risk factor measurements and MI incidence than for a single measurement. Predicted values obtained by linear regression of previous measurements were less well associated with MI incidence.

Repeated measurements enhance precision and reduce misclassification in risk groups, but when prolonged over too many years, the association with disease is attenuated due to selective loss of subjects at high risk. This bias is quantitatively more important than reduction in intraindividual variation in this cohort.

Recommendations

For epidemiological longitudinal studies directed towards detecting the magnitude of an association between biological risk factors and disease, more emphasis should be placed on correction of intraperson variation by using repeated measurements, even where measurements are technically as accurate as possible. This way, random error and the underestimation of the association between risk factors and disease can be reduced. In etiologic studies, with long follow-up periods varying between 10 and 25 years, precision (avoidance of random error) in assessing risk factor levels deserves more consideration.

Repeated measurements should take place in a relatively short period, to avoid bias due to selective loss of high-risk subjects.

Acknowledgements

This study was supported by a grant from the Netherlands Prevention Foundation. The authors would like to thank the men in Zutphen who took part in the repeated surveys, the fieldwork team in Zutphen, especially E.B. Bosschieter, MD, Mrs. A.H. Thomassen-Gysbers, the late Prof. Dr. F.S.P. van Buchem, the principal investigator of the Zutphen Study in the period 1960–1975, and Ms. M. Drijver, MD, for coding the morbidity data.

References

1 Rothman KJ: Modern Epidemiology. Boston, Little, Brown, 1986.
2 Healy MJR: Measuring measuring errors. Stat Med 1989;8:893–906.
3 Keys A: Coronary heart disease in seven countries. Circulation 1970; 41(suppl 1):I1–I211.
4 Kromhout D, Bosschieter EB, Drijver M, de Lezenne Coulander C: Serum cholesterol and 25-year incidence of and mortality from myocardial infarction and cancer. The Zutphen Study. Arch Intern Med 1988;148:1051–1055.
5 Willett W: An overview of issues related to the correction of nondifferential exposure measurement error in epidemiologic studies. Stat Med 1989;8:1031–1040.
6 Tornberg SA, Jakobson KFS, Eklund GA: Stability and validity of a single serum cholesterol measurement in a prospective cohort study. Int J Epidemiol 1988;17:797–803.
7 Thompson SG, Pocock SJ: The variability of serum cholesterol measurements: Implications for screening and monitoring. J Clin Epidemiol 1990;43:783–789.
8 MacMahon S, Peto R, Cutler J, Collins R, Sorlie P, Neaton J, Abbott R, Godwin J, Dyer A, Stamler J: Blood pressure, stroke and coronary heart disease. 1. Prolonged differences in blood pressure: Prospective observational studies corrected for the regression dilution bias. Lancet 1990; 335:765–774.
9 Keys A, Aravanis C, Blackburn H, Buzina R, Djordevic BS, Fidanza F, Karvonen MJ, Kimura N, Menotti A, Mohacek I, Nedeljkovic S, Puddu V, Punsar S, Taylor HL, Van Buchem FSP: Seven Countries: A Multivariate Analysis of Death and Coronary Heart Disease. Cambridge, Harvard University Press, 1980.
10 Abell AA, Levy BB, Brody BB: A simplified method for the estimation of total cholesterol in serum and a demonstration of its specificity. J Biol Chem 1952;195:357–366.
11 Anderson JT, Keys A: Cholesterol in serum and lipoprotein fractions: Its measurement and stability. Clin Chem 1956;2:145–149.
12 Rose GA, Blackburn H: Cardiovascular Survey Methods. Geneva, World Health Organization, 1968.
13 Norusis MJ, SPSS Inc: SPSS/PC+ Base Manual/Advanced Statistics. Chicago, SPSS, 1988.
14 Dixon WJ, Brown MB, Engelman O: BMDP Statistical Software Manual. Berkeley, University of California Press, 1988.
15 Cox DR: Regression models and life tables. J R Stat Soc Ser B 1972;34: 187–220.
16 Blomqvist N: On the relation between change and initial value. J Am Stat Assoc 1977;72:746–749.
17 Snedecor GW, Cochran WG: Statistical Methods. Ames, Iowa State University Press, 1980.

18 Bookstein L, Gidding SS, Donovan M, Smith FA: Day-to-day variability of serum cholesterol levels. Impact on the assessment of risk according to the national cholesterol education program guidelines. Arch Intern Med 1990;150:1653–1657.
19 Reed D, Benfante R: Lipid and lipoprotein predictors of coronary heart disease in elderly men in the Honolulu Program. Ann Epidemiol 1992; 2:29–34.
20 Boerma GJM: Volgens welke klinisch-chemische methoden moeten totaal-cholesterol, HDL-cholesterol en triglyceriden worden gemeten? Hart Bull 1987;suppl 1:35–40.
21 Feskens EJM, Bowles CH, Kromhout D: Intra- and interindividual variability of glucose tolerance in an elderly population. J Clin Epidemiol 1991;44:947–953.
22 Shipley MJ, Pocock SJ, Marmot MG: Does plasma cholesterol concentration predict mortality from coronary heart disease in elderly people? 18 year follow up in the Whitehall study. BMJ 1991;303:89–92.
23 Davis CE: The effect of regression to the mean in epidemiologic and clinical studies. Am J Epidemiol 1976; 104:493–498.
24 Cochran WG: Errors of measurement in statistics. Technometrics 1968;10:637–666.
25 Healy MJR, Goldstein H: Regression to the mean. Ann Hum Biology 1978;5:277–280.
26 de Klerk NH, English DR, Armstrong BK: A review of the effects of random measurement error on relative risk estimates in epidemiological studies. Int J Epidemiol 1989;18: 705–712.
27 Keli SO, Bloemberg B, Kromhout D: Predictive value of repeated systolic blood pressure measurements for stroke risk. Stroke 1992;23:347–351.
28 Kleinbaum DG, Kupper LL, Muller KE: Applied regression analysis and other multivariable methods. Boston, PWS-Kent Publishing, 1988, pp 62–63.

Cardiology 1993;82:100–121

Uri Goldbourt[a]
Shlomit Yaari[b]
Jack H. Medalie[c]

[a] Department of Epidemiology and Preventive Medicine, Sackler School of Medicine, Tel Aviv University, and Neufeld Cardiac Research Institute, Sheba Medical Center, Tel Hashomer, Israel;
[b] Bar-Ilan University Computing Center, Ramat Gan, Israel;
[c] Division of Research, Department of Family Medicine, Case Western Reserve University Medical School, Cleveland, Ohio, USA

Factors Predictive of Long-Term Coronary Heart Disease Mortality among 10,059 Male Israeli Civil Servants and Municipal Employees

A 23-Year Mortality Follow-Up in the Israeli Ischemic Heart Disease Study

Abstract

Over 10,000 male civil servants and municipal employees in Israel, aged 40 years and above, underwent an extensive clinical, biochemical, anthropometric, sociodemographic and psychosocial evaluation in 1963, 1965 and 1968. Follow-up for mortality was continued through 1986. Over 23 years, a number of previously established risk factors for coronary heart disease (CHD) incidence were found to predict mortality. The long-term follow-up assisted in illustrating temporal patterns. A single casual assessment of blood pressure retained high prediction for long-term mortality. Blood lipids, while significantly associated with both coronary and all-cause mortality, exhibited a small contribution to the latter, when compared to hypertension, cigarette smoking habits and diabetes. Weak associations of long-term coronary mortality with the dietary intake patterns of fatty acids, as reported at baseline, were probably fully mediated by the effect of the diet on serum cholesterol. Religious orthodoxy appeared to provide a degree of immunity, part of which was independent of life-style correlates. A number of now well-established associations in cardiovascular epidemiology were first demonstrated, or amplified, in the study. Patterns of ethnic diversity in the risk factor and prevalence rates of CHD persisted, as viewed from the angle of mortality rates, over nearly a quarter of a decade, highlighting the enigma of a migrant country as a cardiovascular melting pot.

Key Words

Prospective studies
Cardiovascular disease
Epidemiology
Multivariate analysis
Ethnic diversity
Religiosity
Immigrants
Nutrition

This paper was written while U.G. was a Visiting Scientist at the Division of Epidemiology and Clinical Application, National Heart, Lung and Blood Institute, National Institutes of Health, Bethesda, Md., USA. Data collection, analysis, the baseline and the two incidence examinations (1963–1968) were part of a collaborative study by NIH, the Ministry of Health, Israel, and the Hadassah Medical Organization, supported by P.L. 480 counterpart funds, Research Agreement No. 375106.

Uri Goldbourt, PhD
Neufeld Cardiac Research Institute
Sheba Medical Center
Tel Hashomer 52621 (Israel)

Introduction

In order to investigate factors related to the development of coronary heart disease (CHD), hypertension and diabetes in a multicultural society engaged in a nationwide process of absorbing mass immigration, the Israeli Ischemic Heart Disease (IIHD) prospective study was initiated in 1961 as a collaborative undertaking of the Israeli Ministry of Health, Hadassah Medical Organization and the US National Institutes of Health.

Earlier impressions had suggested that hospitalizations for CHD in European-born Jews in Israeli hospitals were more frequent than in other population groups; and some evidence has existed to suggest that over the years the influence of life in Israel on differing nutritional and behavioral habits, originally experienced by the different immigrant groups, is likely to bring about changes in cardiovascular disease patterns [1].

Subjects and Methods

Subjects

With the above in mind, and given the well-known socioeconomic and cultural differences between the diverse groups of immigrants, the study participants have been chosen by stratified sampling. The sampling aimed at obtaining approximately equal numbers of study subjects born in five different geographic-cultural Jewish immigrant groups, as well as Jews born in the pre 1948 British Mandate of Palestine. Those invited to participate included 5/7 of all Jewish men, aged 40 and over, born in central Europe (FRG, Czechoslovakia, Hungary, Austria and Switzerland), 1/4 of eastern European born (USSR and Poland, including those born during the short independence period of the Baltic republics), 5/6 of those born in the Balkan countries (southeastern Europe) all of those born in the pre-1948 borders of British-dominated Palestine, mostly within the post-1948 borders of independent Israel, and the mideastern Asian countries as well as all such Jewish male employees born in northern Africa (Morocco, Algeria, Tunisia, Libya and Egypt). A sample of 14,283 men was thus chosen.

After the sampling framework was established, civil and municipal service collaboration secured, and adjustment carried out to include only men working in the seven districts nearest the three examining centers in Haifa, Tel Aviv and Jerusalem, 11,876 tenured civil servants and municipal employees, aged 40 and above, were invited for examination. The logistics and the examination procedures have been described previously [2].

The civil service and municipal work force in Israel consisted of such diverse professions as teachers, social workers, postmen, prison wardens, dock workers, workers in the railway transport system, cleaning personnel, engineers, scientists and administrators. It therefore provided an extensive representation of the different occupations and educational/socioeconomic levels in the country [2].

Of those invited to participate, 10,232 (86.2%) actually participated in the first medical examination in 1963. These individuals also provided diverse sociodemographic and behavioral information, responded to a psychosocial questionnaire, were interviewed for dietary habits [3] and underwent electrocardiographic, biochemical, blood and genetic (blood group) testing [4]. Follow-up was restricted to 10,059 individuals whose age was ascertained as falling within the originally agreed upon age span and who were born within the six predefined geographic areas. Nonrespondents resembled respondents in their age and area-of-birth distribution but tended to be more ill.

On a repeat examination in 1965, in which 95% of invited participants took part, the major characteristics collected at the baseline examination were reassessed. Additional information was collected pertaining to a number of sociological questions, including topics, such as religious education and upbringing, concentration camp experience during the holocaust period in Europe, status incongruity and occupational mobility [5]. In 1968, the third and last follow-up medical and electrocardiographic evaluation of the cohort took place, in which an assessment of spirometric function was added for a portion of the study population. Analysis of the morbidity and mortality experience between 1963 and 1968 confirmed several hypothesized ethnic patterns and illuminated others [6]. It also identified a host of factors predisposing to the incidence of the different manifestation of coronary and other chronic diseases. The results of the 5-year incidence study of myocardial infarction (MI) [6], angina pectoris [7], diabetes [8] and hypertension [9] have been extensively reported. Other examinations, the results of which will not be mentioned here, included an assessment of spirometric function for

43% of the study sample and limited study of fundus examinations and glucose tolerance tests.

Over a period of 23 years (1963–1986), a total of 3,473 deaths were recorded, including 1,098 attributed to CHD as the underlying cause. Analyses undertaken in this paper focus on the characteristics that predisposed these 1,098 individuals to fatal CHD, as compared to counterparts surviving the follow-up period or dying from other causes. Information on mortality was derived from the Israeli Mortality Registry. Practically all deaths are accounted for. The cause of death was documented as CHD on the basis of individual determination by a review panel during the 1970s and according to the International Classification of Diseases, version 9, thereafter. Additional details have been published [10].

Methods of Measurement of Study Variables

The measurement of blood pressure by the examining physician was done in the recumbent position on the right arm, approximately 30–45 min after arrival. Systolic blood pressure (SBP) was measured at the onset of Korotkoff sounds, and diastolic blood pressure (DBP) was measured upon the disappearance of sounds (or muffling if the sound did not disappear). Fifteen to 30 min after the first measurement, blood pressure was recorded again using an identical method. The second measurement of blood pressure is used in the results reported herein.

Measurement of cholesterol and other blood biochemistry parameters were done in a special research laboratory from venous blood drawn in a nonfasting state after electrocardiography, but before a psychosocial and dietary survey. Cholesterol was determined by the Anderson and Keys [11] modification of the Abell method. Cholesterol in the HDL fraction was measured by the method of Burstein and Samaille [12] after precipitation of the low and very low density fractions. Apparently some of the HDL was also precipitated as shown by values consistently lower than those found for Israeli middle-aged men in CDC-controlled samples. However, examination of the relationship of serum HDL cholesterol assessed in 1965 to CHD mortality during the years 1965–1978, as compared to the association of 1963 HDL cholesterol with 1963–1978 CHD mortality, has indicated that the correlation of HDL cholesterol with outcome has not been materially affected [13].

The initial appraisal of the study population for diabetes mellitus [14] was twofold. All men were asked about prior diagnoses of diabetes or insulin injection during the course of medical history taking. A venous blood sample was subsequently taken for blood glucose determination. No further estimates of values definitely below 130 mg/dl were undertaken. For all others, more exact determinations by the Somogyi-Nelson or Hagedorn-Jensen methods were made. The details of classifying individuals as definite, probable and possible diabetics, or as showing abnormal glucose tolerance have been published [14].

Previous heart attack was considered to be present when a participant reported a positive history and the ECG taken at the time of the study showed changes consistent with a previous MI. If the ECG failed to confirm the history, a search was made of the medical summaries and the ECG tracing at the reported time of heart attack. Verification in these cases depended on review of medical summaries, laboratory findings and rereading of old ECGs by two cardiologists.

The diagnosis of MI was based on: (a) definite MI by either an ECG interpretation of probable or possible infarct (classified by the study diagnostic criteria developed from the Minnesota code) or autopsy findings showing definite evidence of recent MI and/or coronary thrombosis, and (b) probable MI, in case of sudden unexpected death occurring within 1 h of onset of symptoms or a composite of clinical, laboratory and ECG findings of ischemic changes in hospitalized subjects [6].

Definite angina was diagnosed using the Rose questionnaire, in which typical presternal pain caused by physical exertion and relieved within 10 min by discontinuation of the effort or by intake of a vasodilating drug considered necessary.

Individual dietary assessments were carried out using a short dietary questionnaire based on a scoring system of frequency × nutrients × amount. Special diets, meal and postprandial consumption patterns were also taken into account. Food models provided visual aids in estimating the quantities consumed. Details of the questionnaire, its administration, its validity examination and quality control have been published [2]. As intake of linolenic acid (a very small quantity relative to that of linoleic acid) was not determined, we approximated the polyunsaturated/saturated fat intake ratio (P/S) by the ratio of linoleic to saturated fat intake.

An assessment of religious orthodoxy was based on three questions pertaining to: (a) education (religious or secular or both); (b) self-definition (orthodox, 'traditional' or secular), and (c) synagogue-going (never, high holidays only, all holidays and the Sabbath, daily). A combined index with five ranks was used to rank the individual.

Psychosocial variables based on models of stress, support and ways of coping were defined on the basis of questions relating to satisfaction, problems and reaction in family and work relationships. During study examination, interobserver variation between examiners was monitored for blood pressure readings, angina assessment in terms of presence and severity, and other diagnoses. The frequency of extreme responses to the psychosocial questionnaire, namely 'very good' and 'very bad' (or 'very unhappy') was also monitored to unearth possible effects of the interviewer's personality on the response by the interviewee. This resulted in retraining and occasionally in the elimination of an examination as being too unreliable (for example, arcus senilis). Additional details concerning the measurement of study variables have been published [15].

Direct age-adjusted rates of CHD and total mortality have been computed in person years according to 10-year age groups taking into acccount the shifting age of the study population over the long follow-up period [16]. A total of 203,892 person-years for follow-up were available for analysis (200,865 for men with serum chemistry examined).

Multivariate analysis of mortality was performed with Cox's life table proportional hazards model [17] with estimators derived from Breslow's [18] modification for tied observations. Exponentiation of the coefficients obtained by the latter methods and of the same coefficients ± 1.645 SE yielded the point estimates and the associated 95% confidence intervals for covariate-adjusted estimated relative 23-year mortality risk associated with given differences in risk factor levels. Life table risks in quintiles of blood pressure, total cholesterol and HDL cholesterol and in cigarette smoking categories were computed (a) by Kaplan-Meier survival estimates and (b) by stratifying proportional hazard estimates on quintiles (or nominal categories) of variables, independent of assumptions regarding the underlying distribution of the risk factor (for details see ref. [19]). The approximate deciles and quintiles of variables tabulated in tables 3 and 5 are listed below (for calculated variables, e.g. percent cholesterol in HDL or nutrient ratios, wherever identical values form the upper bound of a percentile as well the lower bound of the next one, the upper bound implies 'less than' whereas the lower bound implies 'that value or greater'): total cholesterol (mg/dl): <161, 161–176, 177–187, 188–197, 198–206, 207–216, 217–227, 228–241, 242–260 and >260; percent cholesterol in HDL fraction: <11.7, 11.7–13.5 (that is all ratios of at least 11.7 but smaller than 13.5; a ratio could theoretically equal 13.499), 13.5–14.9, 14.9–16.1, 16.1–17.3, 17.3–18.7, 18.7–20.2, 20.2–22.3, 22.3–25.5 and >25.5%; SBP (mm Hg): ⩽113, 114–118, 119–123, 124–127, 128–130, 131–137, 138–140, 141–148, 149–162 and ⩾163 (because of digit preference, the number of subjects in approximate deciles fluctuated between 797 and 1,318); DBP (approximate quintiles, n fluctuating due to major preference for 80 mm Hg): ⩽75 (n = 1,904), 76–78 (n = 1,093), 79–84 (n = 2,963), 85–90 (n = 2,119) and ⩾91 (n = 1,978); dietary variables (quintiles): saturated fat (g/week) intake: <131, 131–173, 173–213, 214–267 and ⩾267; saturated fat intake as percent of total fat: <27.6, 27.6–30.6, 30.6–33.3, 33.3–36.4 and ⩾36.4%; linoleic acid intake (g/week): <82, 82–114, 115–146, 147–192 and ⩾192; linoleic acid intake as a fraction of the saturated fat intake (P/S ratio): <0.47, 0.47–0.62, 0.62–0.76, 0.76–0.96 and ⩾0.96, and oleic acid intake as a fraction of the saturated fat intake (M/S ratio): <0.93, 0.93–1.02, 1.02–1.12, 1.12–1.26 and ⩾1.26.

While the lower and upper limits of the above deciles seldom coincided with biologically meaningful limits, the categorical cutpoints were applied for the evaluation of associations across the whole range of biological and dietary variables examined to long-term mortality.

Results

Over the duration of the 23-year follow-up period (1963–1986), 3,473 (34.5%) of study subjects died. In slightly under one third of these deaths the underlying cause was coded as CHD. The distribution of mortality by cause is shown in table 1. CHD mortality rose with age from 50/10,000 in those entering the study atage 40–44 to 234 in those aged 60 and above at entry.

Since this study was undertaken in strata predefined by six areas of birth, the age-adjusted rates for CHD and all-cause mortality in each of these strata are shown (table 2). The largest group, of men born in the mideastern Asian countries, provided almost 50,000 person-years for follow-up of mortality and experienced the lowest rates of CHD mortality (age adjusted 42.2/10,000 person-years) and the most favorable overall survival over the 23 years of follow-up. The northern-African-born men also exhibited relatively low

Table 1. Mortality over 23 years (1963–1986) according to age at first examination and underlying causes: the IIHD Study

Age at beginning of study, years	Subjects	CHD		CVA	Cancer	Other and unknown causes	Total	
		n	rate[a]				n	rate[a]
40–44	3,317	167	50	33	110	237	547	165
45–49	2,107	180	85	55	116	217	568	270
50–54	2,180	284	130	97	174	280	835	383
55–59	1,492	242	162	92	171	307	812	544
≥60	963	225	234	87	138	261	711	738
Total	10,059	1,098	109	364	709	1,302	3,473	345
Total, %	100.0	31.6		10.5	20.4	37.5		

[a] Per 1,000 persons.
CVA = Cerebrovascular accident.

Table 2. CHD and total mortality, 1963–1986 in six ethnic-geographic strata defined by area of birth

Area of birth	Subjects	Person-years of follow-up	CHD mortality			Total mortality		
			n	rates		n	rates	
				crude	age-adjusted		crude	age-adjusted
Israel[a]	1,431	28,427	172	60.5	60.6	529	186.1	186.0
Eastern Europe	1,928	38,127	258	67.7	57.9	750	196.7	166.9
Balkan countries	1,735	34,777	214	61.5	57.8	653	187.8	174.6
Central Europe	1,374	27,952	156	55.8	55.5	446	159.6	158.6
North Africa	1,219	25,439	107	42.2	51.8	395	155.8	202.5
Mideast	2,372	49,349	191	38.7	42.2	700	141.8	153.6
Total	10,059	203,982	1,098	53.8	53.8	3,473	170.3	170.3

[a] Pre-1948 borders of League of Nations British Mandate on Palestine.
Rates are directly age-adjusted to the overall study age distribution, per 10,000 person-years of follow-up.

CHD mortality, but in contrast to their Asian counterparts suffered the highest overall mortality. The eastern- and southeastern-european-born subjects, who exhibited the highest CHD mortality, along with first- and second-generation Israeli-born persons, exhibited interim rates of total mortality.

The age-adjusted death rates for CHD as well as total mortality per 10,000 person-years of follow-up are shown in table 3 in percentiles of continuous variables established as CHD risk factors. The comparison of such rates does not rule out the effect of intraindividual variability. With this limitation in

Table 3. CHD and total mortality, 1963–1986 by percentiles of risk factors

Deciles	Mortality							
	SBP		DBP[a]		total cholesterol		% cholesterol in HDL	
	CHD	total	CHD	total	CHD	total	CHD	total
1	25	100			36	149	105	228
2	33	129	33	129	35	150	85	199
3	35	136			32	149	64	177
4	36	116	33	128	44	166	58	168
5	44	153			44	142	59	166
6	46	145	44	149	49	156	43	145
7	50	161			57	173	47	155
8	61	190	59	173	68	178	36	156
9	82	217			79	204	25	141
10	128	357	96	266	98	233	27	142

Rates are age adjusted per 10,000 person-years of follow-up.
[a] Approximate quintiles (see methods).

mind, total serum cholesterol, SBP and DBP all displayed a curvilinear relationship with long-term CHD mortality, namely an increase in mortality that occurred predominantly at high levels of these risk factors. CHD mortality in fact showed very little or no rise through SBP levels below 128 mm Hg, DBP levels below 80 mm Hg and cholesterol below 207 mg/dl. At levels exceeding these cutpoints, pronounced stepwise increases in mortality were observed. Data for serum HDL cholesterol levels were available only for two thirds of the study sample. Considerable elevation of long-term CHD mortality was observed in men with a low percentage of cholesterol in HDL, inversely resembling the relationship with the previously mentioned risk factors.

With regard to all-cause mortality, a somewhat erratic pattern was apparent below the median of the SBP distribution, whereas above it risk increased stepwise with SBP elevation. For DBP, levels of 80 mm Hg and above increased the risk in a stepwise fashion. Increased all-cause mortality was associated with low percentages of cholesterol appearing in HDL, but this association dwindled at increasingly larger percentages. In general, the gradient of increasing risk of fatal CHD with increasing blood pressure (in particular SBP) exceeded that associated with increasing total, or decreasing HDL, serum cholesterol levels. Similar patterns were seen for all-cause mortality. For example, men whose initial casual SBP measure at the time of the 1963 examination was below 114 mm Hg, irrespective of whether that measurement was labile as reflected by two successive measurements or repeated measurements over the years, experienced a risk of merely 25 and 100/10,000 person-years, for CHD and all-cause mortality, respectively. Conversely, an initial SBP above 163, subject to the above limitations and irrespective of whichever therapy might have been instituted, implied rates as high as 128 and 357/10,000 person-years for the above noted end points, respectively.

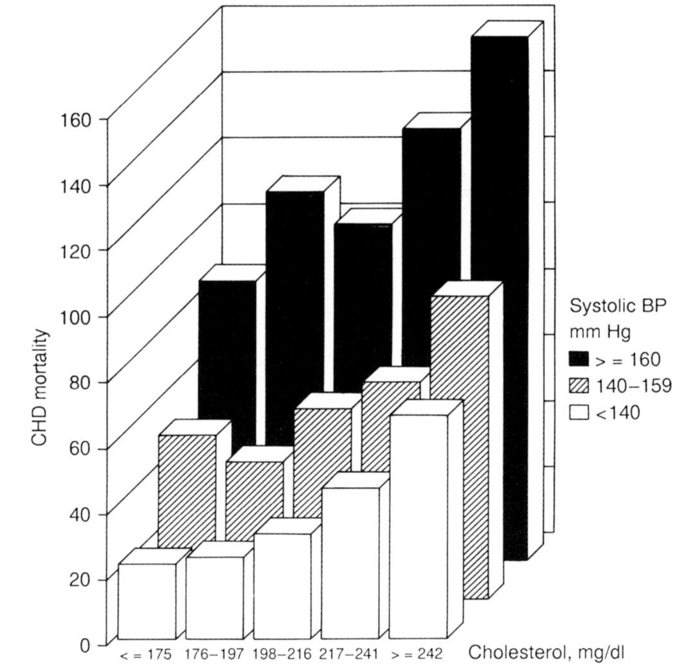

	< = 175	176–197	198–216	217–241	> = 242
> = 160	85	112	102	131	159
140–159	50	42	58	66	92
<140	23	25	32	46	68

Fig. 1. Age-adjusted 23-year CHD mortality in quintiles of cholesterol, divided according to 'normal' (<140), 'borderline' (140–159) and elevated (≥160 mm Hg) levels of SBP. Rates per 10,000 person-years.

In order to better understand the interplay of some of these risk factors, and since the large number of person-years at risk allowed bivariate analysis with significant power, we examined several simultaneous associations of two risk factors with subsequent mortality. Among the different analyses done, we report two significant findings. Figure 1 shows that irrespective of hypertensive status, crudely expressed in wide groups of 'normal' (SBP ≤140 mm Hg), 'borderline' (140–159) and 'systolic hypertension' (≥160), a curvilinear increase in fatal CHD risk with increasing total cholesterol was observed. Likewise, regardless of the quintile of serum cholesterol, the risk of long-term CHD mortality increased markedly with increasing SBP. The latter increase was consistent and large.

Figure 2 examines the joint effects of total cholesterol and HDL cholesterol on long-term CHD mortality. Elevated HDL cholesterol implied reduced risk of fatal CHD at serum concentrations of total cholesterol both below and above 200 mg/dl.

Table 4 demonstrates the major predictive power of cigarette smoking habits in 1963 for fatal CHD and total mortality over the ensuing 23 years. A near doubling in the risk of CHD and total mortality was observed in study subjects who had smoked 20 cigarettes or more per day, in 1963, as compared to those who had never smoked. A

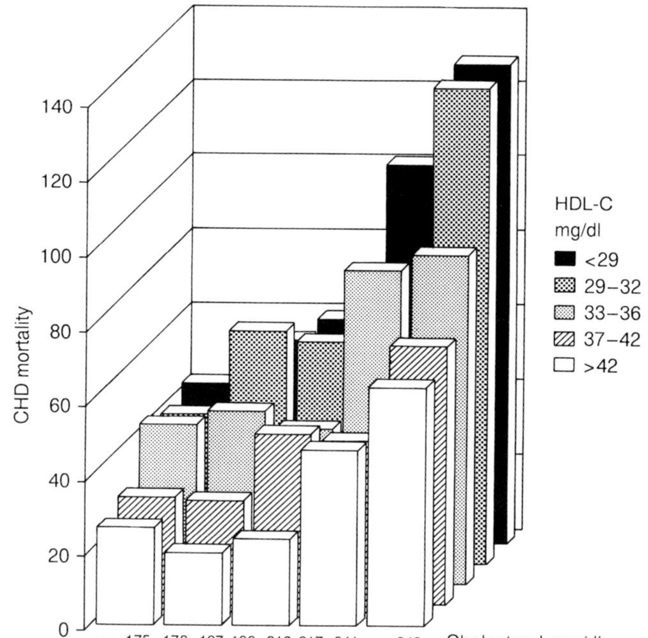

HDL-C mg/dl	< = 175	176–197	198–216	217–241	> = 242
<29	42.4	54.2	59.8	101.2	128.4
29–32	39.6	62.1	59.3	73.8	127.4
33–36	42.5	46.1	41.4	84	88.2
37–42	28.7	27.8	45.7	43.9	69.4
>42	26.2	19.4	23.2	47	63.8

Cholesterol, mg/dl

Fig. 2. Age-adjusted 23-year CHD mortality in quintiles of HDL cholesterol by quintiles of total cholesterol. Rates per 10,000 person-years.

major difference between diabetics at baseline (whether previously known or newly diagnosed) in terms of subsequent mortality, both from CHD and all causes, is evident. As many as 311 (65%) of the 479 subjects diagnosed as diabetics in 1963 (ascertainment methods have been previously described in detail [8] died over the 23-year follow-up period, about twice the death rate for the overall cohort; 41% of these deaths were attributed to CHD.

Two distinct features of a long-term follow-up of Israeli Jews are the ability to relate the degree of religious orthodoxy, as assessed on the basis of education and religious practice, to long-term health and the availability of pertinent health variables and long-term follow-up for a relatively large group of men who survived the horrors of concentration camps in Europe (primarily in Germany and Poland) during World War II. The highly orthodox religious groups of the study experienced significantly reduced CHD death rates (38/10,000 person-years) as compared with the other participants, and this difference was similarly seen in the all-cause mortality rates. Reexamination of these differences by area of birth did not eliminate the observed differences. No differences in survival between 690 concentration camp survivors during the holocaust and 2,322 counterparts immigrating from Europe after 1938 were found, the death

Table 4. CHD and total mortality by smoking habits, diabetes and religious orthodoxy, 1963–1986

Smoking status	Cases	Mortality	
		CHD	all causes
Never smoked	3,158	40	129
Quit smoking	1,677	51	151
Current smokers			
1–10 cigs/day	1,498	47	153
11–20 cigs/day	1,634	60	193
>20 cigs/day	1,944	70	237
Cigar and pipe	141	72	164
Diabetics	479	149	357
Nondiabetics	9,580	50	162
Religious orthodoxy			
Most orthodox	2,103	38	135
Orthodox	1,528	51	162
'Traditional'	1,782	51	162
Secular	2,085	54	166
'Nonbelievers'	1,747	61	168

Rates are age-adjusted per 10,000 person-years of follow-up.

rates and CHD-free survival patterns being virtually equal in both groups.

Table 5 presents findings relating intake of saturated and polyunsaturated fatty acids, as estimated in 1963 on the basis of questionnaires recording information on the 'usual food intake over a period of 1 week' to subsequent 23-year CHD and total mortality. While both the absolute intake of saturated as well as polyunsaturated fat were inversely related to CHD mortality, adjustment for the total fat consumed produces distinctly different findings. Calculated *percent of fat* derived from saturated fat was related to *increased CHD mortality* whereas the ratio of linoleic fat intake to saturated fat intake (P/S ratio) was inversely related to this end point. The latter findings were virtually identical when linoleic acid intake was calculated as a percentage of total fat intake (results not shown). The absolute intake of oleic acid, the major monounsaturated fatty acid, revealed no association with CHD mortality; on the other hand, the ratio of oleic acid intake to saturated fat (M/S ratio) intake closely resembled the P/S ratio in showing an inverse relationship to CHD mortality. With respect to total mortality, high unadjusted intakes of saturated and linoleic fat, as well as of oleic acid (not shown) were related to reduced mortality. After adjustment for total fat intake, no such relationshp remained for saturated fat intake. Conversely, when intake of linoleic and oleic acid was examined relative to saturated fat intake, men in the lowest quintile of consumption still showed somewhat increased total mortality (statistically significant only for the P/S ratio). The association of CHD mortality with high percent of fat consumed as saturated, and with low P/S and M/S ratios, were all statistically significant by a test for linear trend. However, the variation in age-adjusted fatal CHD rates was small; the rates varied in the range of 48–58/10,000 person-years, regardless of the fat intake pattern.

Multivariate Analysis of Mortality

Table 6 presents the covariate-adjusted relative risks (RRs) of the 23-year CHD and all-cause mortality associated standard increments in the continuous risk factors as well as with the presence of a number of disorders at study entry. All the RRs are significantly different from 1 ($p \leq 0.002$). With respect to CHD mortality, a 20 mm Hg increment in SBP at the single casual reading taken in visit 1 was associated with a 1.5-fold increased risk of fatal CHD, as estimated from a life table model accounting for the intercorrelation between risk factors. A 10-year increment in age was independently associated with a doubling in the risk of fatal CHD. Increments (or, for HDL cholesterol, decreases) of approximately 1 SD for the examined lipids were only associ-

Table 5. CHD and total mortality by estimated intake of dietary nutrients 1963–1986

Quintiles	Saturated fat intake				Linoleic acid		P/S ratio		M/S ratio	
	absolute		as percent of fat							
	CHD	total	CHD	total	CHD	total	CHD	total	CHD	total
1	61	192	48	170	63	199	57	188	58	183
2	56	173	55	163	55	169	58	174	57	166
3	54	172	53	165	54	166	53	166	53	168
4	50	155	57	173	51	166	52	157	53	162
5	49	161	58	182	47	155	49	169	50	175

P/S = Ratio of linoleic (polyunsaturated) to saturated fatty acid intake, M/S = ratio of oleic (monounsaturated) to saturated fatty acid intake.

Rates are age-adjusted by 10,000 person-years of follow-up.

Table 6. Multivariate analysis of CHD and all-cause mortality, 1963–1986

Risk factor	CHD mortality (723 deaths)		All-cause mortality (2,233 deaths)[a]		Difference associated with RR
	RR	95% CI	RR	95% CI	
Age	2.02	1.80–2.26	2.31	2.07–2.35	10 years
SBP	1.50	1.42–1.59	1.40	1.35–1.45	20 mm Hg
Smoking	1.35	1.22–1.50	1.42	1.34–1.51	11–20 cigs/day vs. never
Cholesterol					
Total	1.29	1.20–1.39	1.07	1.03–1.12	40 mg/dl
HDL	1.36	1.25–1.49	1.09	1.04–1.14	10 mg/dl (reduced)
Diabetes	2.73	2.16–3.44	2.16	1.86–2.50	previously known or newly diagnosed
History of MI	5.26	3.94–7.01	2.61	2.08–3.27	verified vs. none or refuted
Angina pectoris	1.48	1.11–1.98	1.44	1.19–1.73	definite vs. none

RR = Relative risk; CI = confidence interval.

[a] RRs for total mortality adjusted also for presence of initial malignant disease.

ated with excess fatal CHD risk of approximately one third. Cigarette smoking, without knowledge of subsequent change, preserved an independent predictive power of a similar magnitude. A major factor in determining the risk of fatal CHD was the diagnosis of diabetes and prior MI at the time of the baseline examination.

The excess mortality risk associated with increased age, high blood pressure and cigarette smoking carries over from CHD to all-cause death, and the same holds true for diabetes. However, the increased risk of CHD mortality associated with cholesterol and its HDL fraction is considerably diminished for all-cause death. A reanalysis excluding 557

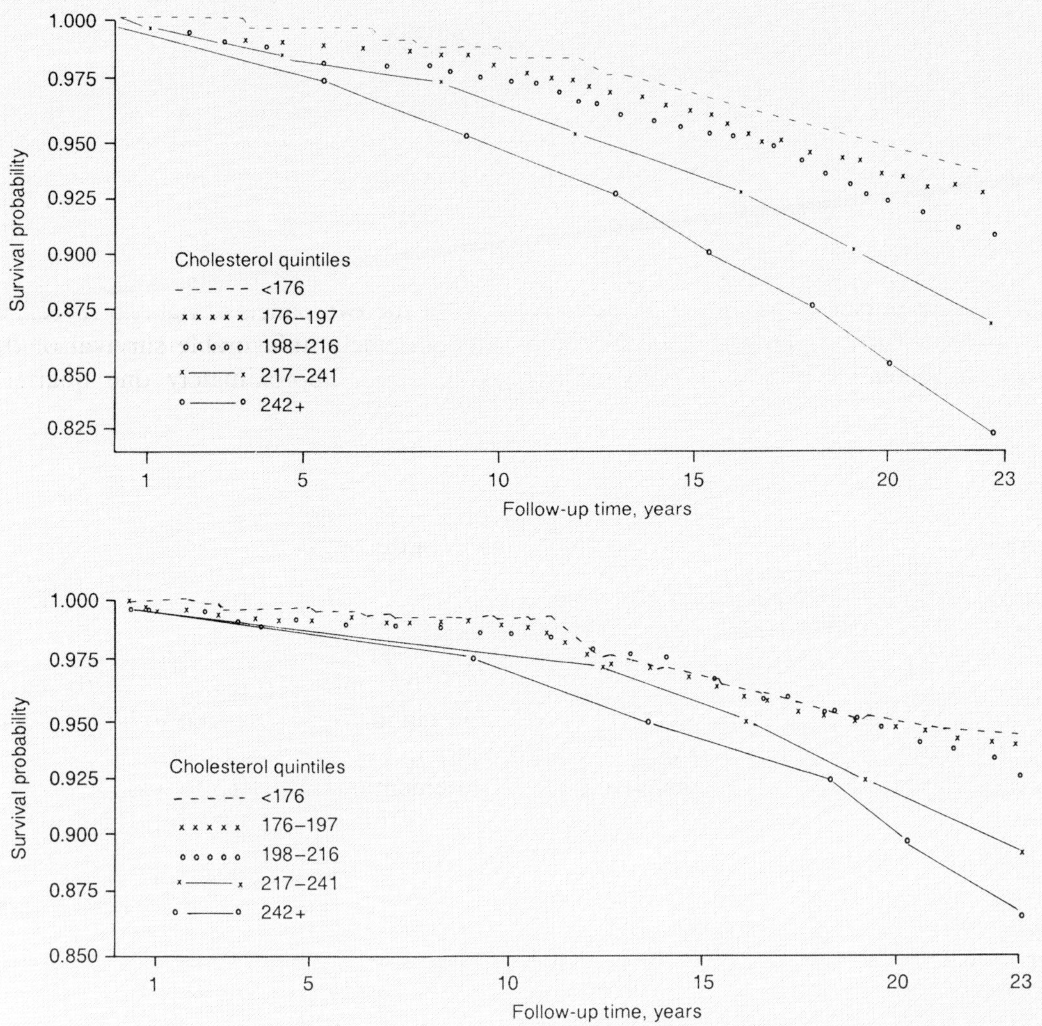

Fig. 3. Kaplan-Meier (above) and adjusted (below) life table curves by quintiles of total serum cholesterol: 23-year CHD mortality adjusted by age, SBP, HDL cholesterol, smoking habits and angina.

study subjects with coronary disease and diabetes (among whom 173 died of CHD) and therefore based on only 550 CHD deaths made negligible differences in the RR estimates.

Upon examination of the above-mentioned dietary fat intake-CHD mortality associations in multivariate analysis, the adjustment of each of these associations for serum cholesterol eliminated the statistical significance, and the age- and cholesterol-adjusted RRs have approached 1 (data not tabulated).

Temporal Patterns of Associations with Risk Factors

During the first 5 years of the study (1963–1968), subjects maintained close contact with the examining centers. This contact ceased after the incidence examinations were finished. Any favorable effect of study participation in terms of closely monitored health status probably faded over the years. Concurrently, an increasing trend of rapidly declining CHD mortality became evident in Israel beginning about a decade into the mortality follow-up of this cohort and proceeded through the follow-up period [21]. In view of these factors and the possibility of diminishing associations with baseline examination results, the life table experience of the cohort by categories of risk factors is of special interest.

Figure 3 shows the Kaplan-Meier curves of fatal CHD-free survival in quintiles of 1963 serum cholesterol levels (upper panel) and the covariate-adjusted survival (lower panel). It indicates a limited propensity of men in the 60% lowest baseline cholesterol levels towards increased CHD death which, however, does not persist independently as suggested by the lower panel, when the intercorrelation with the other risk factors is accounted for. About 10 years into the study, a clear disadvantage of cholesterol levels ≥217 mg/dl as well as an independent risk associated with levels >240 mg/dl became apparent. For HDL cholesterol (fig. 4) a similar adjustment for total cholesterol and the other factors emphasizes an advantageous risk status in terms of CHD mortality for those men in the highest 40% HDL cholesterol range, a difference that emerges after 12 years and appears to become prominent after 20 years.

The Kaplan-Meier survival curves for SBP (fig. 5) exhibited an immediate increase in fatal CHD among the group with baseline casual readings of ≥149 mm Hg in 1963. The mean SBP of these men was 166 mm Hg. This group subsequently exhibited the worst CHD mortality experience of any subgroup of the study sample, with the exception of diabetics and those who had already entered the study with CHD or malignant disease. A comparison of the two panels of figure 5 reveals that the extremely unfavorable survival of these individuals, approximately one quarter of whom died of CHD, included an important contribution of old age and other correlates of blood pressure. Individuals with an initially high SBP reading remained, however, independently of age and other risk factors, at materially increased risk of dying from CHD. The covariate-adjusted risks of fatal CHD over time for the other SBP quintiles (with the respective means of 111, 122, 131 and 142 mm Hg in 1963) appeared to be equally distanced, but none approached that of the aforementioned group. Similar analyses by initial DBP (not shown) yielded similar results, except that the risk differences between the top quintile (DBP >90 mm Hg in 1963, with mean DBP = 101) and the other DBP categories were somewhat less pronounced than in the case of SBP and the two bottom categories (DBP ≤75 and 76–78 mm Hg, with mean DBPs of 70 and 77 mm Hg, respectively) showed almost no difference in either crude or unadjusted fatal CHD risk over the 23-year follow-up period.

Despite a statistically significant independent prediction ($p = 0.04$) of fatal CHD by the Quetelet index ('body mass index'; QI), the average excess independent risk associated with 3.3 kg/m^2 (approximately 1 SD), estimated from this follow-up, is only 6.5%. Life table analysis of fatal CHD by initial QI (not shown) did indicate survival advantage for

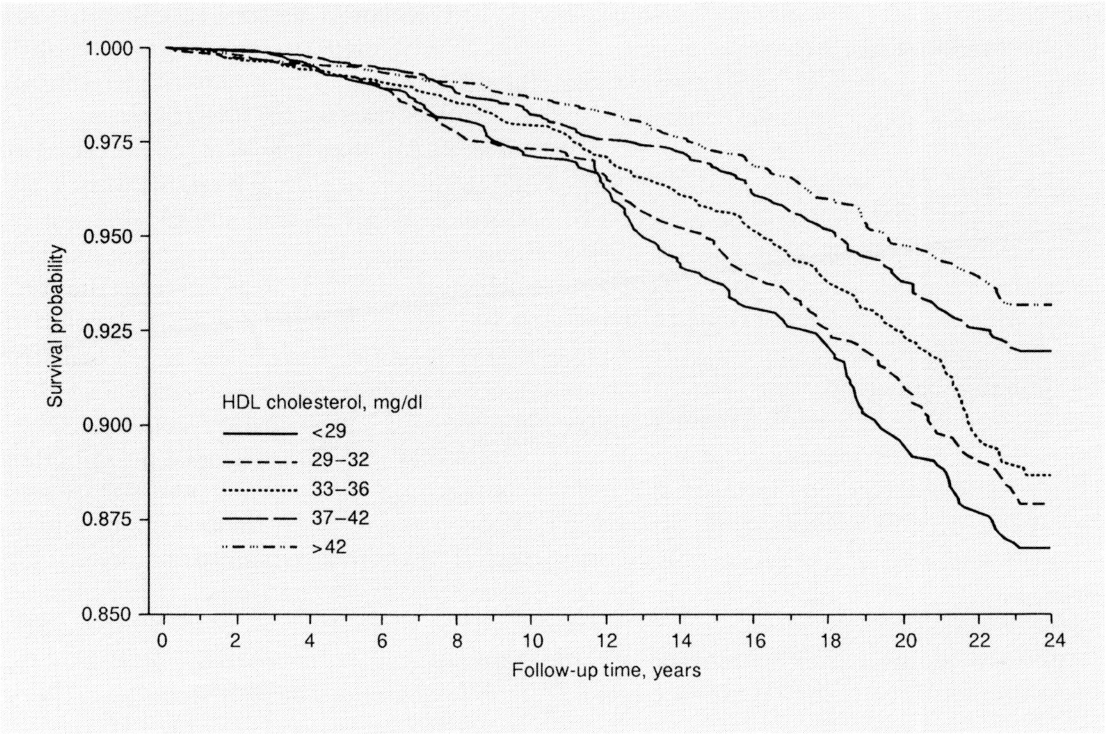

Fig. 4. Adjusted life table curves by quintiles of HDL cholesterol: 23-year CHD mortality adjusted by age, SBP, total cholesterol, cigarette smoking, diabetes, QI and presence of CHD at baseline.

the leanest quintile (mean QI = 21.1 kg/m^2) and a disadvantage for the heaviest one (mean QI = 30.3). The covariate-adjusted life table, however, eliminated most of these differences, indicating that beyond the knowledge of serum cholesterol, blood pressure and diabetes and CHD status of the individual, differences in relative weight added very little to predictive power.

A comparison of Kaplan-Meier and adjusted life table for diabetics and nondiabetics entering the study (not shown) indicated a gradual deterioration of the diabetic group in terms of CHD mortality. The cumulative CHD mortality rate of 34.4% for the diabetics was 'reduced' to an adjusted figure of 21.7% when all other differences between diabetics and nondiabetics were controlled for, in comparison to CHD death rates of 11.5 (unadjusted) and 9.5% (adjusted) for the nondiabetics. Diabetes mellitus therefore remained the strongest precursor of long-term coronary death in CHD-free individuals in this cohort of middle-aged and elderly men.

Cigarette smoking at the beginning of the study (fig. 6) provided significant prediction of the likelihood of fatal CHD. The upper panel shows an unfavorable CHD mortality pattern for former smokers as well as for those who had reported smoking more than 10 cigarettes/day. Adjustment for age and other risk factors suggested, however (lower panel), that, given their older age, quitters have in fact fared as well as those who had begun the study

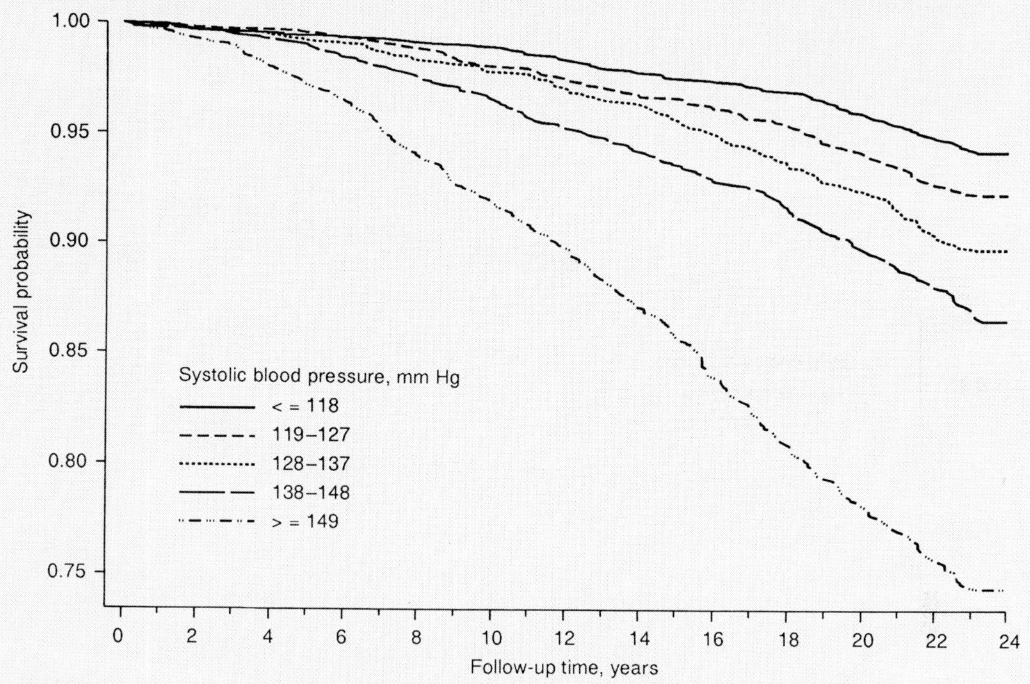

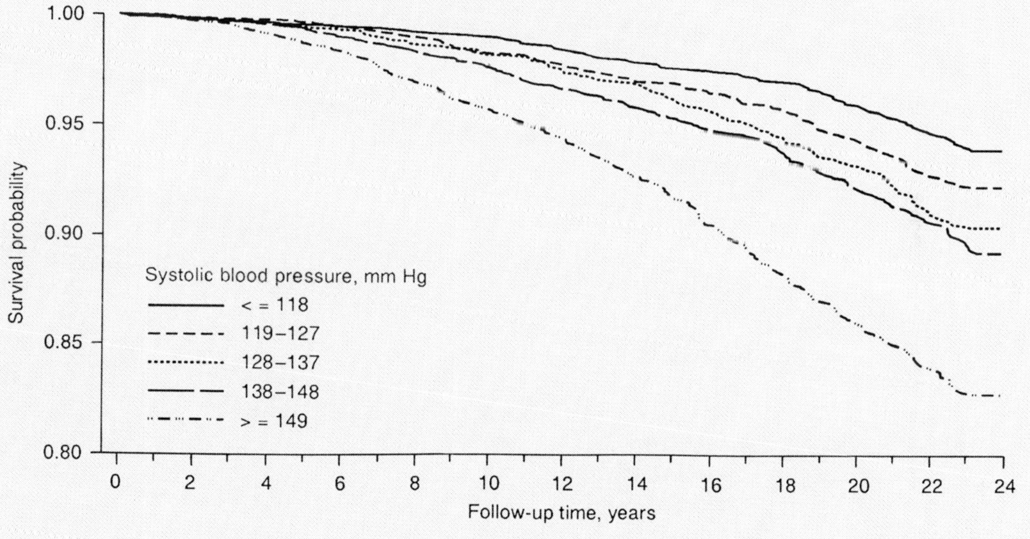

Fig. 5. Kaplan-Meier (above) and adjusted (below) life table curves by quintiles of SBP: 23-year CHD mortality adjusted by age, total cholesterol, cigarette smoking, diabetes, QI and the presence of CHD at baseline.

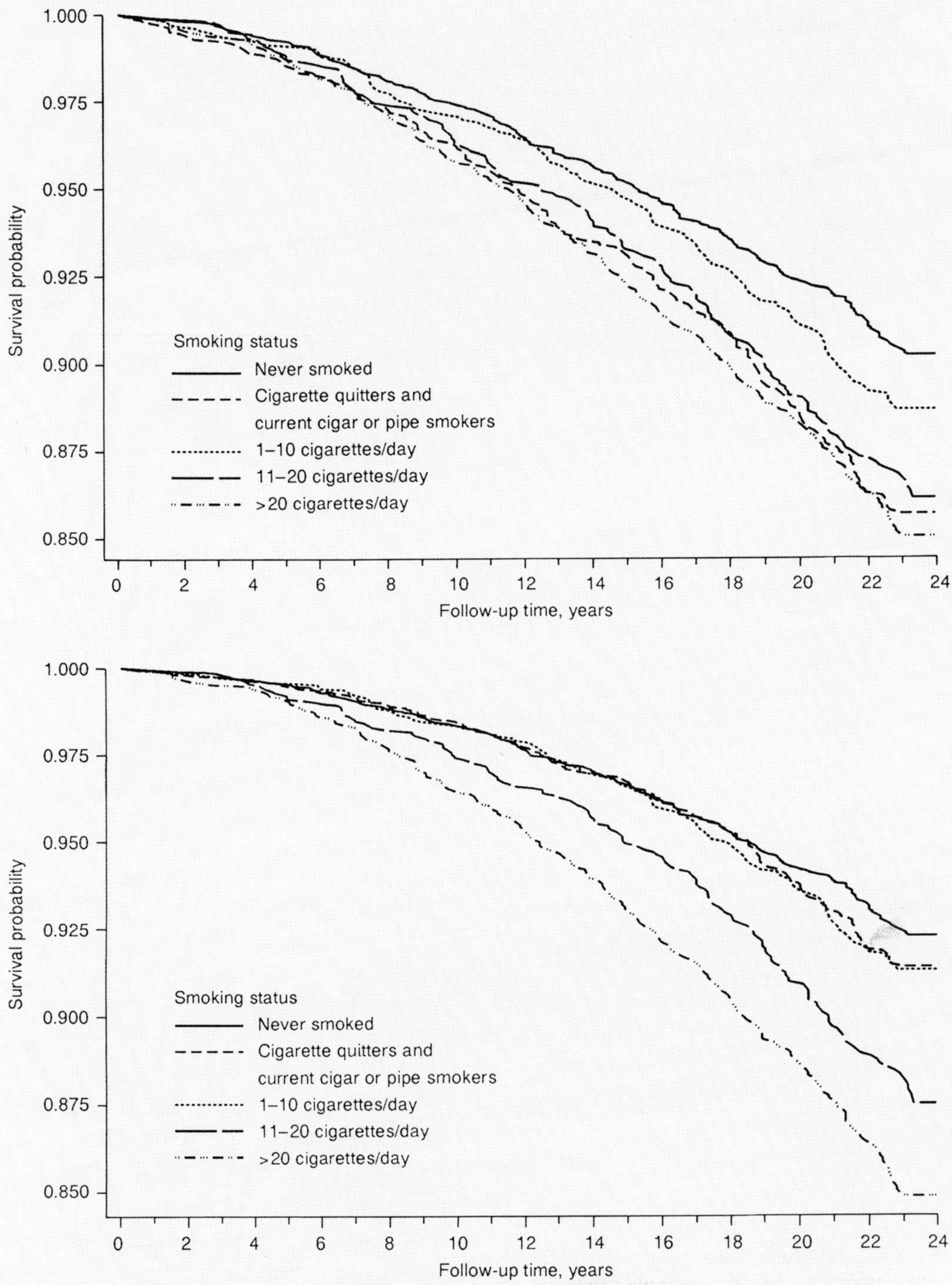
1.000
0.975
0.950
0.925
0.900
0.875
0.850
Survival probability
Smoking status
Never smoked
Cigarette quitters and
current cigar or pipe smokers
1–10 cigarettes/day
11–20 cigarettes/day
>20 cigarettes/day
0
2
4
6
8
10
12
14
16
18
20
22
24
Follow-up time, years
1.000
0.975
0.950
0.925
0.900
0.875
0.850
Survival probability
Smoking status
Never smoked
Cigarette quitters and
current cigar or pipe smokers
1–10 cigarettes/day
11–20 cigarettes/day
>20 cigarettes/day
0
2
4
6
8
10
12
14
16
18
20
22
24
Follow-up time, years

6

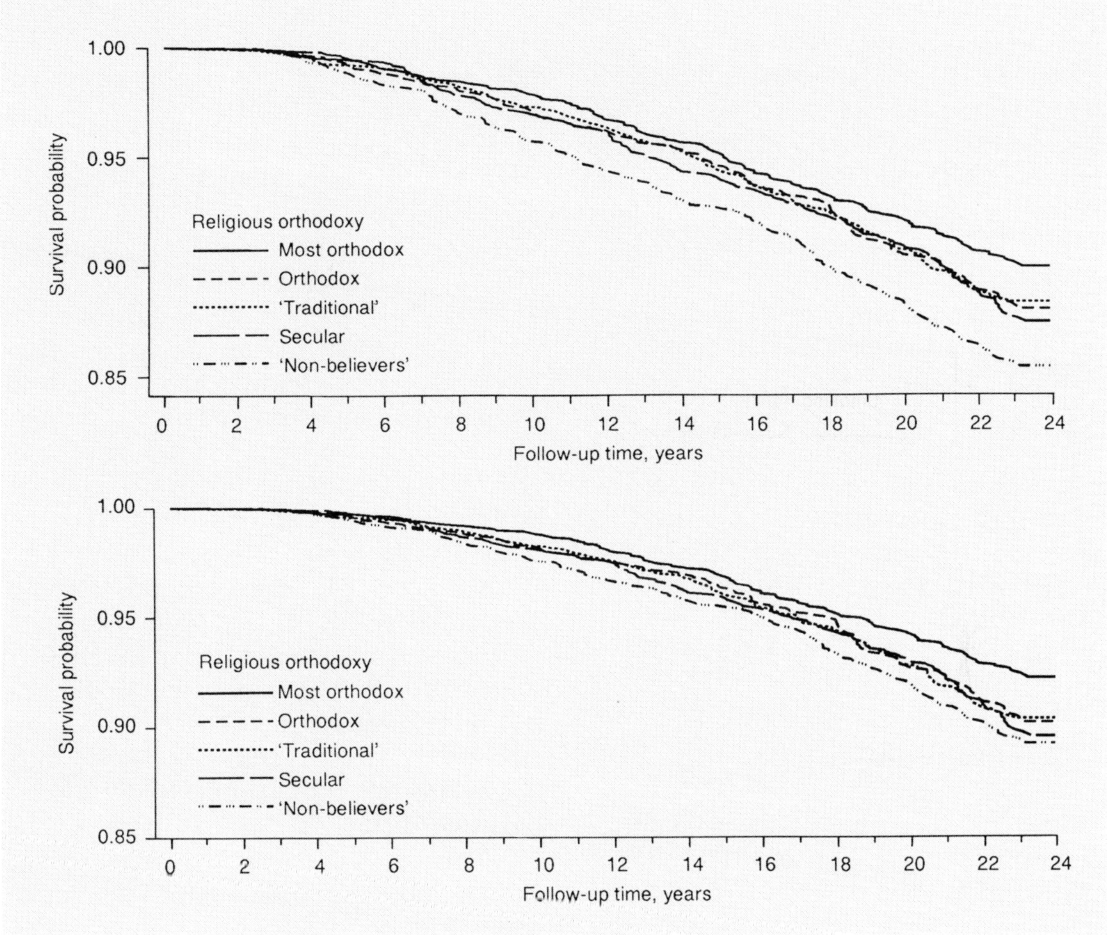

Fig. 7. Kaplan-Meier (above) and adjusted (below) life table curves by degree of religious orthodoxy: 23-year CHD mortality adjusted by age, SBP, total cholesterol, cigarette smoking, diabetes, QI and presence of CHD at baseline.

smoking 10 cigarettes or less a day and only slightly worse than 'never smokers' in 1963. A marked disadvantage for heavier smokers in 1963 (regardless of our inability to incorporate information concerning subsequent changes in smoking habits) was accentuated by the adjustment. Thus cigarette smoking is the only risk factor demonstrating a considerably increased association with CHD mortality as a result of adjustment for covariates.

Finally, we further investigated the previously noted tendency of the 'most orthodox' Jewish study participants to exhibit reduced long-term CHD mortality. Figure 7 shows a

Fig. 6. Kaplan-Meier (above) and adjusted (below) life table curves by cigarette smoking categories: 23-year CHD mortality adjusted by age, SBP, total cholesterol, diabetes, QI and presence of CHD at baseline.

crude cumulative CHD mortality of 10.0% for this group, as opposed to the significantly increased CHD mortality among those defining themselves as fully detached from religious practices and beliefs ('nonbelievers', cumulative CHD mortality = 14.5%). In comparing the upper (unadjusted) to the lower (adjusted) life table curves for these groups, it appeared that the difference between the above-mentioned most secular group and the more traditional (or less secular) groups is, in most part, accounted for by a more unfavorable CHD risk profile for the former. However, a statistically significant advantage of at least 20% reduced fatal CHD risk persisted for the 2,103 'most orthodox' (adjusted CHD mortality, 7.7%) over the years, as compared to the other groups (adjusted cumulative CHD mortality rates between 9.7 and 10.7%).

Discussion

The results of the univariate und multivariate analyses of the 23-year CHD mortality reflect and strengthen previous findings obtained from this study when the baseline prevalence and subsequently the incidence of MI and angina pectoris over the first 5 years (1963–1968) of follow-up [6, 7, 20] were analyzed. These analyses found the highest prevalence and 5-year incidence rates of MI among subjects born in central and eastern Europe, followed by men born in Israel and the Balkan countries, including Greece. Immigrants from the Balkan countries also exhibited very high rates of definite angina pectoris [7]. Subjects born in mideastern and northern African Arab countries exhibited the lowest prevalence and incidence rates of these clinical manifestations of CHD. The variation in CHD incidence by birth place could not be fully explained by differences between these groups, in the distribution of age and CHD risk factors, even though these factors did offer considerable discriminating power between eventual cases (MI through 1968) and noncases [7, 20]. A finding of particular interest is the persistence, despite cultural and economic changes and a pronounced trend of declining CHD mortality in Israel beginning about a decade after the baseline examination of the study [21], of the ethnic differences in CHD rates. After 5 years, the highest incidence rates of new MI were observed in Jews originating from eastern and central Europe, ahead of the second (and very few third) generation Israeli born and immigrants from the Balkan countries. The long-term CHD mortality rates have now somewhat changed the picture in that the Israeli born have shown slightly higher rates of fatal CHD than counterparts from eastern and central Europe. We do not have the information that would permit us to attribute this change to changing case-fatality rates, changing incidence rates beyond the first 5 years of the study, or combinations thereof. The overall rates of fatal CHD in Israel still suggest that migrants from Europe show the highest rates [22]. Immigrants from the neighboring Arab countries, in particular those from mideastern Asian countries, exhibited the lowest long-term rates of fatal CHD, in keeping with their earlier reduced short-term incidence rates. However, only among the latter are these low rates concomitant with reduced overall mortality. The African-born participants in our cohort, in fact, suffered the highest all-cause mortality. This is also consistent with national mortality data [22]. The explanation of these-findings in immigrants from northern African countries remains obscure. Although a detailed analysis by individual country of birth is outside the scope of the present paper, we have shown elsewhere that CHD incidence among immigrants from mideastern Asian countries varied considerably by country of birth, with Yemen, and to a smaller degree Iraq and Iran, exhibiting relatively lower rates [6]. Findings for fatal CHD in the total population of Israel have been similar [23]. To what extent the continuing variability of the toll which CHD exerts on the Israeli population varies by origin due to genetic effects, vis-à-vis sociocultural and other environmental differences, has been a topic of great interest over the years [23]. In our study, certain differences in the 5-year incidence of MI and angina were observed in association with the ABO blood groups with a possible very high risk (a post hoc finding) associated with the A1BJka– combination of the ABO and Kidd phenotypes [4]. The genetic aspects of CHD in Israel have been reviewed recently [24].

An unusual feature of this study was the inclusion of psychosocial factors in addition to the biological ones, in the assessment of risk for developing MI, angina and hypertension over a 5-year period [7]. Anxiety and family problems were equally and synergistically important with cigarette smoking, high blood pressure and serum cholesterol shown to be associated with the development of angina [7]. The same psychosocial markers did not exert a statistically independent effect on CHD mortality over 23 years of study follow-up.

Unlike the early incidence analysis [7, 20] which relied on logistic regression estimation of 'net' odds ratios associated with diabetes mellitus, increased age and blood pressure or serum lipid elevations, and previous analyses that examined mortality in this cohort [10], we have now examined mortality with considerable attention to the aging of this cohort. Person-year exposure has, in the present analysis, been accounted for in 'moving' groups of actual age, rather than by age at the time of study entry. While our data do not provide the opportunity of biennially updating the risk profile of each individual using repeated risk factor measurements, in the manner of the Framingham Heart Study, it provides a better expression of the disease burden continuously increasing as the population has been aging. It should be kept in mind, however, that a cohort effect may also exist; a person aged 60–69 during the years 1963–1972, for example, may not incur an identical risk of fatal CHD to that of a counterpart, even with a similar risk profile, aging into this group in the late 1970s. Another caveat is the remoteness of the original measurement from distant events such as death 20 years later. In fact, given these limitations, it is surprising how powerful a single casual blood pressure reading, taken in 1963, has proven to be for predicting mortality over almost a quarter of a century (table 3, fig. 1) – despite the problems arising from digit preference, the well-established large intraindividual variability of blood pressure and the resultant regression to the mean. Recently, it has been pointed out that because of the 'regression-dilution effect' the association of blood pressure with subsequent fatal CHD is in all likelihood considerably stronger than the extent suggested by the analysis of CHD rates in blood pressure assessed by a single reading [25]. The availability of repeated blood pressure measurements in the IIHD study in 1965 and 1968 renders possible a more refined analysis than we have undertaken in the framework of this more general analysis. Another source of differences between an analysis of 5-year incidence and 23-year mortality rates is that the antecedents of fatal CHD may not exactly duplicate, nominally or in terms of their qualitative association with CHD, factors which predispose to fatal and nonfatal CHD. Of the long-term epidemiological studies of precursors associated with CHD, only the Framingham Heart Study can address this issue specifically, by virtue of the repeated incidence examinations, so that incidence, rather than (or in addition to) mortality can be analyzed. All the other studies are limited to the examination of the association between baseline risk factor levels and fatal end points.

A number of significant points for the understanding of the roles of the risk factors for CHD have emerged from this and previous analyses in this very large cohort which has provided more than 200,000 person-years for follow-up.

First, the measurement of serum HDL cholesterol during the baseline prevalence examination in 1963 has rendered possible the original identification of the *prospective* inverse association of cholesterol in that lipoprotein fraction with incident MI [26]. We have also reported the first finding of an inverse dose-response association of cigarette smoking and HDL cholesterol [27] and one of the first indications that, regardless of the total serum cholesterol level, having a low cholesterol concentration in HDL exposed middle-aged and elderly men to significantly increased fatal CHD risk [28]. The latter point is further accentuated (fig. 2) in the present analysis, highlighting the combination of increased serum cholesterol with a reduced fraction of the latter in HDL as one carrying extremely high risk. Second, we have found an increased risk for fatal CHD risk in association with elevated blood pressure at all levels of total serum cholesterol. The risk is approximately fourfold for men with initial SBP $\geq$ 160 mm Hg and total cholesterol $<$200 mg/dl, as compared to counterparts in the same total cholesterol bracket, whose baseline casual SBP readings were $<$140 mm Hg. These findings accentuate the inadvisability of designing extensive intervention strategies concentrating exclusively on total serum cholesterol [29]. These shortcomings have become apparent, and modified population- and individual-based lipid intervention strategies [30] may be forthcoming.

In examining the life table experience of the study sample, it is clear that despite continuously declining rates of CHD mortality in Israel [21], the relative hazards of cigarette smoking, diabetes mellitus, high total (and LDL) cholesterol and blood pressure and low HDL cholesterol have been maintained. However, the curves for quintiles of the different continuous variables, as well as that of men who never smoked in comparison to counterparts who had quit cigarettes, did not diverge as clearly over the first decade after the initial examination as they do subsequently. Perhaps the early differences in CHD incidence [7, 26] require longer to influence the rates of fatal CHD.

The decline in CHD mortality in Israel did not coincide with any known changes in smoking habits, known decreases in fat intake (in a country that enjoys one of the highest known ratios of polyunsaturated/saturated fatty acid intake in the western world [31]), or with changing blood cholesterol levels [21]. Therefore, declining rates of fatal CHD in Israel, during 1970–1985, may have reflected major improvements in diagnostic procedures, the advent of coronary care units which were started in the late 1960s, and the availability, in most areas, of increasingly efficient means of transporting individuals with typical chest pain to these units and other improvements in acute care management. These changes notwithstanding the risk factors for CHD have continued to provide sizable predictive power for long-term fatal CHD and, given the caveats discussed above, may in fact possess stronger discriminating power than these or other data indicate.

The association of the 'major risk factors', to long-term CHD mortality is consistent with those previously found with the 5-year incidence findings. As far as dietary intake is concerned, no association between saturated fat intake in 1963 and incident CHD through

1968 was apparent in our analysis of the latter [26]. Only limited evidence for an inverse association between the P/S fatty acid intake ratio and MI incidence has been observed for CHD incidence. Such associations are notoriously difficult to unravel, presumably because of the large intraindividual variability, which renders any single estimate of dieatry intake unreliable. Long-term mortality analyses, such as those from the Chicago Western Electric study, have provided limited independent predictive power associated with polyunsaturated fat and cholesterol intake [32, 33]. The analysis of long-term CHD mortality in the IIHD Study indicated possible protection associated with increased proportions of dietary fat ingested as either poly- or monounsaturates. The absence of an independent contribution of dietary variables to CHD risk is of secondary importance given the established effect of diets replacing saturated fat by other fat types on the blood lipid profile. The relative merit of diets differing in their total fat and types of fat remains a major topic of clinical investigation [34].

A special aspect, concomitant with the marked ethnic, cultural and occupational diversity of the cohort under study, is the heterogeneity of these men in the degree of deviation from the almost uniform highly orthodox way of life in their country of origin. With the possible exception of Germany and Austria, the overwhelming majority of parents of our study subjects have preserved all the 'Mitzvot' (practical day-to-day religious rules), in particular those related to Kosher foods, prayers and the following to the letter of rules of the Sabbath. The collapse of the traditional Jewish way of life in the diaspora at the end of the 19th and, in particular, the beginning of the 20th century resulted in varying degrees of 'secularization' of the younger generation, among some of whom immigration to Palestine prior to World War II may have further removed them from religious traditions. The distribution by the degree of religious orthodoxy differed considerably by country of origin. Our multivariate analysis of fatal CHD events according to religiosity indicated a small 'independent' advantage, in terms of the probability of dying from CHD and all causes, enjoyed by 20% of the sample who reported themselves as the most orthodox. In another study in Israel, albeit of a case-control nature, MI odds ratios as high as 4.2 and 7.3 for men, after adjustment for 'conventional' risk factors, have been estimated for secular persons relative to religious ones among Jewish residents of Jerusalem [35]. The association between frequent religious attendance and better health has been reviewed [36]. The largest North American prospective study of this nature [37] found low rates of fatal CHD among California Mormon high priests and their wives, the 1980–1987 standardized mortality ratios for cardiovascular disease among them slightly exceeding one half of those for the overall white population. Since the results of our study suggest that differences in the distribution of blood pressure, serum cholesterol, cigarette smoking, prevalence of diabetes and prior CHD do not eliminate the advantage of the highly orthodox group, additional research is suggested to examine other associated life habits and other possible environmental sources of variability.

In this paper, we have given limited information on predictive factors for total mortality. Total mortality differed appreciably by area of birth, by baseline blood pressure and cigarette smoking and was twice as high in diabetics as in nondiabetics. It showed little, albeit statistically significant, association with total and HDL cholesterol. Total mortality was significantly reduced in the 'most orthodox' individuals and somewhat elevated in persons consuming very low proportions of their fat as linoleic acids. The latter associations require further study.

In summary, we have reported herein the findings of a 23-year CHD mortality follow-up in a large cohort of male civil servants and municipal employees. Over the years, members of this cohort have aged from a mean age of about 49 years to age 72 (range: 63 to approximately 88 years, with a few exceptions), have in general retired from the civil and municipal service, and have lived in years of political and social turmoil. This cohort has also dwelled under increasing relative affluence and with the advent of vastly improved medical care. In this population, risk factors for CHD, previously identified as being of predictive importance in our incidence study and by others, continued to play a central role in exposing these men to mortality. The ethnic diversity in terms of CHD risk and the risk associated with diabetes mellitus and increased levels of the known coronary risk factors persisted in this cohort despite the previously described developments in their occupational and social environment. The IIHD study has significantly increased the degree of understanding of CHD and its correlates in a migrant society that has, over the years, acted as a melting pot for a large number of diverse population groups.

Acknowledgments

We are greatly indebted to Joan Pinsky for her assistance in preparing the illustrations for this paper and to Michael Wolz in facilitating data transfer, handling and analysis.

References

1 Toor M, Katchalsky A, Agmon J, Allalouf D: Atherosclerosis and related factors in immigrants in Israel. Circulation 1960;22:265–279.
2 Groen JJ, Medalie JH, Neufeld HN, et al: An epidemiologic investigation of hypertension and ischemic heart disease within a defined segment of the adult male population of Israel. Isr J Med Sci 1968;4:177–194.
3 Balogh M, Medalie JH, Smith HA, Groen JJ: The development of a dietary questionnaire for an ischemic heart disease survey. Isr J Med Sci 1968;4:195–203.
4 Medalie JH, Levine C, Papier C, Goldbourt U, Dreyfuss F, Oron D, Neufeld HN, Riss E: Blood groups, myocardial infarction and angina pectoris among 10,000 males. N Engl J Med 1971;285:1348–1353.
5 Shagmar L, Leslau A, Medalie JH (eds): Methodological Research Manual, Israel Ischemic Heart Disease Study, vol VI: Sociological Aspects. Jerusalem, 1967.
6 Medalie JH, Kahn HA, Neufeld HN, Riss E, Goldbourt U: Five-year myocardial infarction. I. Prevalence, incidence and mortality. J Chronic Dis 1973;26:63–84.
7 Medalie JH, Goldbourt U: Angina pectoris among 10,000 men. II. Psychosocial and other risk factors as evidenced by a multivariate analysis of a five-year incidence study. Am J Med 1976;60:910–921.
8 Medalie JH, Papier C, Goldbourt U, Herman JB: The major factors in the development of diabetes mellitus among 10,000 adult men. Arch Intern Med 1975;135:811–817.
9 Kahn HA, Medalie JH, Neufeld HN, Riss E, Goldbourt U: The incidence of hypertension and associated factors: The Israeli Ischemic Heart Disease Study. Am Heart J 1972;84:171–182.
10 Goldbourt U, Holtzman E, Neufeld HN: Total and high density lipoprotein cholesterol in the serum and risk of mortality: Evidence of a threshold effect. Br Med J 1985;290:1239–1243.
11 Anderson JT, Keys A: Cholesterol in serum lipoprotein fraction. Clin Chem 1956;21:145–159.
12 Burstein M, Samaille J: Sur un dosage rapide du cholésterol lié aux alpha- et aux bêta-lipoprotéines du sérum. Clin Chim Acta 1960;5:609–611.
13 Goldbourt U, Yaari S, Cohen-Mandelzweig L, Neufeld HN: High-density lipoprotein cholesterol: Correlation with biochemical, anthropometric, behavioral, and clinical parameters in 6,500 Israeli men. Prev Med 1986;15:569–581.
14 Herman JB, Medalie JH, Kahn HA, et al: Diabetes incidence: A two-year follow-up of 10,000 men in a survey of ischemic heart disease in Israel. Diabetes 1967;19:938–943.

15 Medalie JH, Kahn HA, Neufeld HN, Riss E, Groen JJ: Physician's Fact Book. Selected Measurements on 10,000 Israeli Males. Jerusalem, Central Press, 1968.
16 Yaari S, Goldbourt U: A SAS program for evaluating person years of risk in cohort studies. Comp Biol Med 1989;19:353–359.
17 Cox DR: Regression models and life-tables. J R Stat Soc Ser B 1972; 34:187–202.
18 Breslow NE: Analysis of survival data under the proportional hazards model. Int Stat Rev 1975;43:45–58.
19 Goldbourt U, Yaari S: Cholesterol and coronary heart disease mortality. A 23-year follow-up study of 9902 men in Israel. Arteriosclerosis 1990;10:512–519.
20 Goldbourt U, Medalie JH, Neufeld HN: Clinical myocardial infarction over a five-year period. III. A multivariate analysis of incidence. The Israeli Ischemic Heart Disease Study. J Chronic Dis 1975;28:217–237.
21 Goldbourt U, Neufeld HN: Trends in coronary heart disease mortality and related factors. Cardiology 1985;72:63–74.
22 Central Bureau of Statistics and Ministry of Health: Causes of Death 1983–1986. Special Series No. 869. Jerusalem, Central Bureau of Statistics, 1990.
23 Peritz E, Dreyfus F, Halevi HS, Schmelz UO: Mortality of Jews in Israel 1950–1967. Special Series No. 409. Jerusalem, Central Bureau of Statistics, 1973.
24 Goldbourt U: Genetic aspects of coronary heart disease in Israel; in Bonné-Tamir B, Adam A (eds): Genetic Diversity among Jews. New York, Oxford University Press, 1992, pp 377–394.
25 MacMahon S, Peto R, Cutler J, Collins R, Sorlie P, Neaton J, Abbot R, Godwin J, Dyer A, Stamler J: Blood pressure, stroke, and coronary heart disease. 1. Prolonged differences in blood pressure: Prospective observational studies corrected for the regression dilution bias. Lancet 1990; 335:765–774.
26 Medalie JH, Kahn HA, Neufeld HN, Riss E, Goldbourt U: Five-year myocardial infarction incidence. II. Association of single variables by age and birthplace. J Chronic Dis 1973;36:329–349.
27 Goldbourt U, Medalie JH: High density lipoprotein cholesterol and coronary heart disease incidence. Am J Epidemiol 1979;109:296–308.
28 Goldbourt U: The ratio of high-density to total cholesterol: Long-term prognosis in men with and without coronary heart disease and in different ages, ethnic origins and serum cholesterol categories; in Widhalm K, Naito HK (eds): Recent Aspects in Diagnosis and Treatment of Lipoprotein Disorders: Impact on Prevention of Atherosclerotic Diseases. Prog Clin Biol Res. New York, Liss, 1988, vol 255, pp 37–48.
29 Report of the National Cholesterol Education Program Expert Panel on Detection, Evaluation and Treatment of High Blood Cholesterol in Adults. Arch Intern Med 1988;148: 36–69.
30 National Institutes of Health Consensus Development Conference: Triglyceride, High Density Lipoprotein, and Coronary Heart Disease. Panel statement. Bethesda, February 28, 1992.
31 Kaufmann NA, Dennis BH, Heiss G, Friedlander Y, Kark JD, Stein Y: Comparison of nutrient intakes of selected populations in the United States and Israel: The Lipid Research Clinics prevalence study. Am J Clin Nutr 1986;43:604–620.
32 Shekelle RB, Shryock AM, Paul O, Lepper M, Stamler J, Liu S, Raynor WJ: Diet, serum cholesterol, and death from coronary heart disease. The Western Electric Study. N Engl J Med 1981;304:65–70.
33 Shekelle RB, Stamler J: Dietary cholesterol and ischaemic heart disease. Lancet 1989;i:1177–1179.
34 Berry EM, Eisenberg S, Haratz D, Friedlander Y, Norman Y, Kaufmann NA, Stein Y: Effects of diets rich in monounsaturated fatty acids on plasma lipoproteins – The Jerusalem Nutrition Study: High MUFAs vs high PUFAs. Am J Clin Nutr 1991;53:899–907.
35 Friedlander Y, Kark JD, Stein Y: Religious orthodoxy and myocardial infarction in Jerusalem – A case control study. Int J Cardiol 1986;10: 33–41.
36 Levin JS, Vanderpool HY: Is frequent religious attendance really conducive to better health? Toward an epidemiology of religion. Soc Sci Med 1987;24:589–600
37 Enstrom JE: Religious practices and cancer mortality among active California Mormons. J Natl Cancer Inst 1989;81:1807–1814.

Cardiology 1993;82:122–136

Black-White Comparisons of 20-Year Coronary Heart Disease Mortality in the Evans County Heart Study

Curtis G. Hames[a]
Kathryn Rose[b]
Marilyn Knowles[b]
C.E. Davis[c]
Herman A. Tyroler[b]

[a] Evans County Heart Research, Claxton, Ga., USA, and Departments of
[b] Epidemiology and
[c] Biostatistics, UNC, Chapel Hill, N.C., USA

Key Words
Blacks
Mortality
Epidemiology
Evans County

Abstract

The Evans County Heart Study was initiated to compare the epidemiology of coronary heart disease (CHD) of black and white residents of a southeastern United States community. The study was prompted by a clinical observation that CHD appeared to be less frequent in black than white patients, despite a much higher prevalence and greater severity of hypertension in blacks. A total community-based, closed cohort study was initiated with a prevalence survey in 1960–1962. The 20-year follow-up experience of the Evans County cohort confirmed the importance of the major risk factors of serum cholesterol, smoking and blood pressure for middle age and older, black and white men and women for CHD, cardiovascular disease (CVD) and all-cause mortality. Black men were less likely to die of CHD than white men, and this relative protectiveness increased upon adjusting for risk factors and also increased with age. The relative protectiveness of black men for CHD mortality was also reflected in lower black than white CVD and all-cause mortality in men 65 years of age and older.

Introduction and Rationale

The Evans County Heart Study was initiated to compare the epidemiology of coronary heart disease (CHD) of blacks and whites in a southeastern United States community.

The study was prompted by the observation in the medical practice of one of us (C.G.H.) located in Claxton, Evans County, Ga., USA, that black residents, despite an inordinately high prevalence of hypertension, appeared to have a lower 'incidence' of CHD

H.A. Tyroler, MD
Department of Epidemiology
CB No. 7400, McGavran-Greenberg Hall
University of North Carolina
Chapel Hill, NC 27599-7400 (USA)

[1]. Recognizing the strong potential for selection bias among patients seen in a medical practice, a closed cohort epidemiologic study of community residents was designed and carried out by C.G.H., collaboratively with members of the Department of Epidemiology of the School of Public Health, University of North Carolina at Chapel Hill, under the direction of John Cassel [2].

The conceptual basis of the study was the belief that the modern epidemic of CHD, as for all epidemics, was rooted in the social organization of the population; that societal history, norms and values determined health-relevant behavior and influenced physicochemical/biologic exposures of the population in its local environment [3].

At the time of design of the Evans County Study, the diet-cholesterol-heart disease hypothesis was the leading candidate explanation for the ascending limb of the CHD epidemic, then observed in the US and other industrialized nations [4]. At the societal level, populations in industrialized, urbanized, affluent communities were noted to have earlier onset of the epidemic and higher levels of CHD compared to those nonindustrialized, rural and poor. Evans County, in 1960, was a rural, biracial, nonindustrialized, relatively poor community. At the same time in the early 1960s, the concept of 'risk factors' was being explicated, and formally named as such, at the level of study of individuals; also by this time, the three major CHD risk factors – serum cholesterol, blood pressure and smoking – had been identified.

In the design of the Evans County Study, a large number of biomedical characteristics of individuals as well as social attributes were considered as potential determinants of CHD in the community's black and white residents. Additional characteristics (such as lipoprotein fractions at the individual level [5] and group support at the social level [6]) were added to the assessment of cohort members as theoretic and methodological developments took place over time. For the purposes of this paper, however, analyses are focused on the 20-year risk of death among men and women characterized by their race and social status and their levels of the three major risk factors of serum cholesterol, blood pressure and smoking. The a priori expectation was that, if differences in the frequency of CHD or its risk functions were observed between blacks and whites, much of the difference would be attributable to social and behavioral characteristics.

Materials and Methods

A synopsis of the design, methods and materials of the Evans County Heart Study is presented here, as a detailed description of the study has been published previously [7]. In 1960–1962, all noninstitutionalized residents of Evans County aged 40 years and older and a 50% sample of those 15–39 years old were recruited to participate in an epidemiologic, closed community-based cohort study of CHD; 92% participated. This included 94% of white men, 90% of black men, 92% of white women and 94% of black women. The present analyses are based on all black and white participants aged 40 and older at entry. Overall, CHD and cardiovascular disease (CVD) mortality were the outcomes of interest. The underlying cause of death as coded on the death certificate was accepted. Deaths coded as 410–413 in the ICD-8 (International Classification of Diseases, eighth revision) were classified as CHD deaths, and deaths coded as 390–456 in the ICD-8 were classified as CVD deaths.

Participants, at entry, were given complete physical examinations and were interviewed to obtain baseline medical and demographic data. Data ascertained at baseline that were used in the present study include the following: race/socioeconomic status (SES), cholesterol, systolic blood pressure, diastolic blood pressure, height and weight and calculated body mass index (BMI), smoking history, prevalence of CHD and of ECG abnormalities.

One blood pressure measurement was taken on the left arm, with the participant seated. Serum cholesterol was determined using the Abell-Kendall method. BMI

Table 1. Vital status of the cohort: men

	Examined 1960–1962		Status as of May 1, 1980					
			deceased		living		unknown	
	n	%	n	%	n	%	n	%
Total cohort	3,102	100	1,025	33.0	2,013	64.9	64	2.0
All ages ≥40	2,216	100	953	43.0	1,213	54.7	50	2.3
All males ≥40	1,059	100	531	50.1	505	47.7	23	2.2
Males 40–64	866	100	362	41.8	482	55.7	22	2.5
Black	322	100	137	42.5	166	51.5	19	5.9
White	544	100	225	41.3	316	58.1	3	0.6
LSS	294	100	136	46.3	155	52.7	3	1.0
HSS	250	100	89	35.6	161	64.4	0	0.0
Males ≥65	193	100	169	87.6	23	11.9	1	0.5
Black	74	100	63	85.1	10	13.5	1	1.4
White	119	100	106	89.9	13	10.9	9	0.0
LSS	53	100	50	94.3	3	5.7	0	0.0
HSS	66	100	56	84.8	10	15.2	0	0.0

LSS = Low social status; HSS = high social status.

was calculated as weight (kilograms) divided by height squared (meters). CHD prevalence at baseline was defined as a history of angina pectoris, or a history and/or electrocardiographic evidence of old myocardial infarction. Smoking history was ascertained, and participants classified as to whether they were current, ever or never smokers. In instances where there were missing data for the above continuous risk factors, the average levels obtained for the same sex-race, social status and age group (age was categorized into 5-year intervals) were substituted.

SES was measured using the McGuire-White [8] index of social status. This scale is a weighted combination of information based on the head of household's occupation, source of income and education. Stratification into low or high SES was based on the median value for white men. Those having values above the median were classified as low SES, and those having values equal to or less than the median were classified as high SES. Approximately 95% of blacks fell above the median value for white males. Thus, blacks were not stratified by SES. For the purposes of the present study, a three-level variable consisting of all blacks, low and high SES whites was created to represent race/SES.

The distribution of risk factors by race, SES, age and sex was described. Cumulative mortality rates and mortality hazard rates were computed and compared by race, SES, age and sex. Multivariate proportional hazard models of time to death were computed separately for men and women with race, SES, age, blood pressure, serum cholesterol, smoking history, body mass index, prevalence of CHD and Minnesota-coded ECG abnormalities at baseline as predictor variables. Models were computed for deaths attributed to CHD, for all causes and for deaths attributed to all CVD.

Interaction effects of race with each of the risk factors were tested, and none were found to be statistically significant, i.e. there is no evidence of a race by risk factor interaction. Final models as presented omitted interaction terms.

The distribution of the cohort stratified by race/SES, gender and age groups at baseline entry into the study in 1960–1962 and participants' vital status as of May 30, 1980, is set out in table 1 for men and in table 2 for women. Of the 2,216 examinees aged ≥40+, the vital status was known for all but 50 (2.3%) of individuals. Ascertainment of vital status is complete for all but 3 white males, 6 white females, 20 black males and 21 black females; i.e. for 99.9% of white males, 99.9% of white females, 95% of black

Table 2. Vital status of the cohort: women

	Examined 1960–1962		Status as of May 1, 1980					
			deceased		living		unknown	
	n	%	n	%	n	%	n	%
Total cohort	3,102	100	1,025	33.0	2,013	64.9	64	2.0
All ages ⩾40	2,216	100	953	43.0	1,213	54.7	50	2.3
All females ⩾40	1,157	100	422	36.5	708	61.2	27	2.3
Females 40–64	919	100	245	26.7	653	71.0	21	2.3
Black	376	100	120	31.9	241	64.1	15	1.3
White	543	100	125	23.0	412	75.9	6	1.1
LSS	274	100	67	24.5	201	73.3	6	2.2
HSS	269	100	58	21.6	211	78.4	0	0.0
Females ⩾65	238	100	177	74.4	55	23.1	6	2.5
Black	94	100	70	74.5	18	19.1	6	6.4
White	144	100	107	74.3	37	25.7	0	0.0
LSS	64	100	49	76.6	15	23.4	0	0.0
HSS	80	100	58	72.5	22	27.5	0	0.0

LSS = Low social status; HSS = high social status.

males and 95.5% of black females. The proportion with vital status unknown was higher for middle-aged black than white males (5.9 vs. 0.6%) and higher for older black than white females (6.4 vs. 0%). The cumulative risk of dying was similar for blacks and whites among middle-aged males (43 vs. 41%), older males (85 vs. 89%) and older females (75 vs. 74%); in contrast, the cumulative risk of death was greater for middle-aged black than white females (32 vs. 23%). Among whites in each age, gender stratum, the cumulative risk of death was greater for lower than higher social status individuals (for middle-aged males, the comparative rates were 46 vs. 36%; for older males 94 vs. 85%; for middle-aged women 25 vs. 22% and for older women 77 vs. 73%).

Results

The analyses and the results which follow were organized to present comparisons among the race/SES-gender groups in the following: (1) the prevalence of risk factors and of CHD at baseline; (2) mortality rates for CHD, all CVD and all causes, and (3) the risk functions for mortality associated with serum cholesterol, blood pressure and smoking.

Prevalence of Risk Factors and CHD

The mean levels of risk factors at intake (1960–1962) by race/SES and gender within age groups are set out in figure 1. Associations of the mean levels of the risk factors to the index of race/SES are apparent, however these associations are not the same for each of the risk factors. Considering the demographic scale as reflecting increasing SES from blacks to low to high status whites, systolic and diastolic blood pressure each vary strongly and inversely with race/SES; a similar strong association of race/SES with cigarette smoking is present among middle-aged males. There is no association of cigarette smoking with race/SES among women. By contrast, the association of serum cholesterol levels is strongly

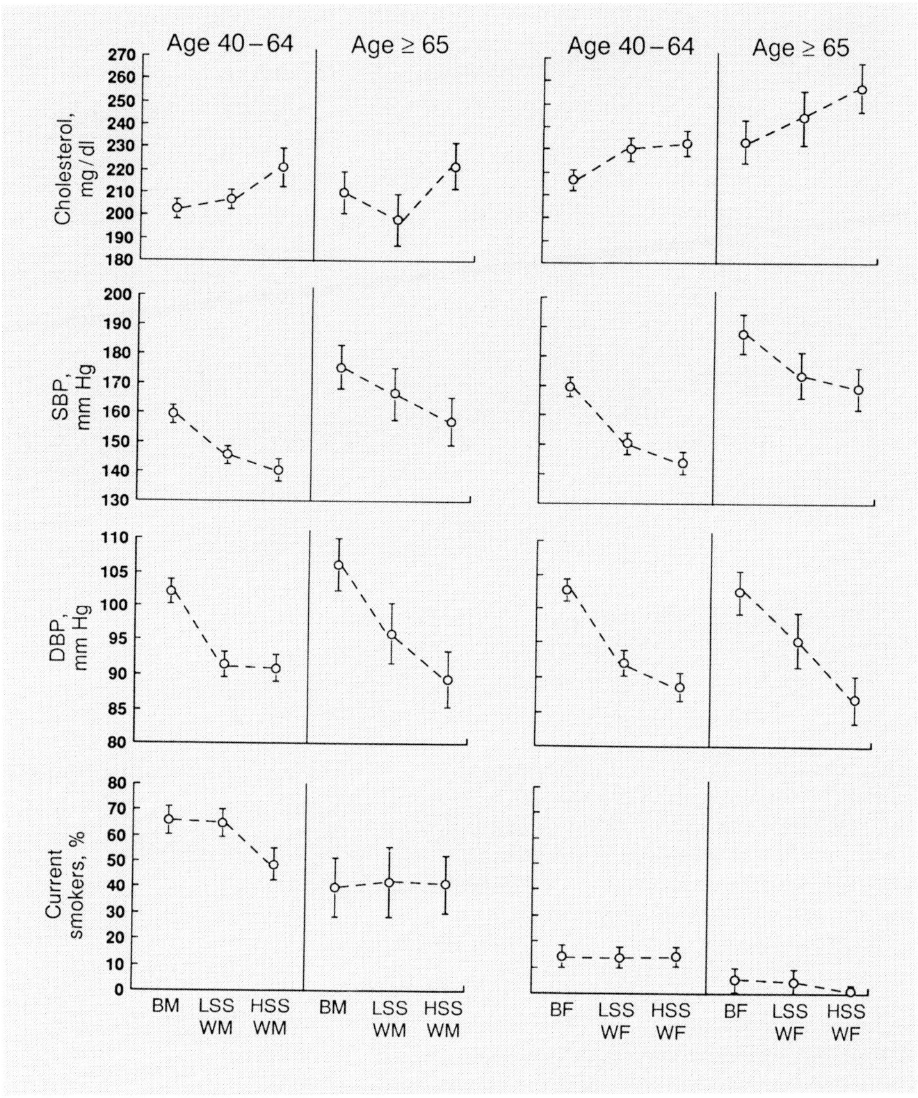

Fig. 1. Mean risk factor levels (± 2 SEM) at intake in 1960 by index of social status and sex within age groups. SBP/DBP = Systolic/diastolic blood pressure; BM/WM = black/white male; LSS/HSS = low/high social status; BF/WF = black/white female.

positive, i.e. the higher the social status the higher the serum cholesterol level for both younger and older males and females.

The association of BMI with race/SES is presented in figure 2. Although not one of the major CHD risk factors, BMI is generally strongly associated with them and is presented for that reason. Its association with race/SES is different among the age-gender groups. The association is positive among younger males, paralleling the association of cholesterol with race/SES, but markedly dif-

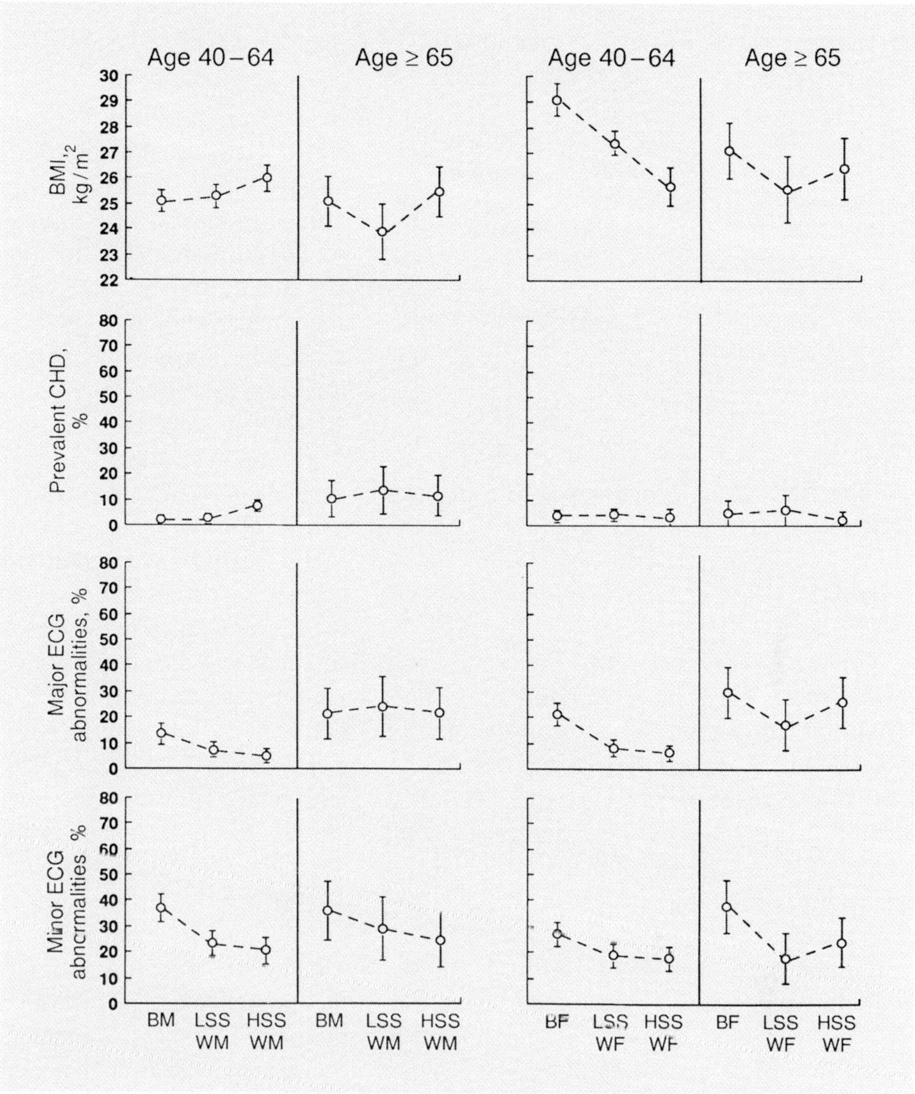

Fig. 2. Prevalence of risk factors at intake in 1960 by index of social status and sex within age groups. For abbreviations, see legend to figure 1.

ferent from the association of blood pressure and smoking; in both older males and females, the association is U-shaped. Among younger women, the association of BMI with SES is strongly negative, paralleling the findings for systolic and diastolic blood pressure and differing from the association of cholesterol with race/SES.

As presented in figure 2, the prevalence of CHD varies directly with race/SES in younger males and inversely in older females. The prevalence of ECG abnormalities varies

Table 3. Black/white cumulative mortality risk ratios by gender and age groups

	Men		Women	
	40–64	≥65	40–64	≥65
Blacks/all whites	1.03	0.96	1.39	1.00
Blacks/LSS whites	0.92	0.90	1.30	0.97
Blacks/HSS whites	1.19	1.00	1.48	1.02

LSS = Low social status; HSS = high social status.

strongly and inversely with race/SES in most of the age-gender strata.

Mortality Rates for CHD, All CVD and All Causes

The black/white cumulative mortality risk ratios for all deaths by gender and age groups are presented in table 3, based on the data of tables 1 and 2. The only sizeable departures from unity are the elevations in the risk ratios for middle-aged black women compared to all white, white lower and white higher social status women with values of 1.39, 1.30 and 1.48, respectively.

On comparison of blacks with all whites, white lower or white higher social status, the black/white cumulative mortality risk ratio is higher in middle-aged than older age groups, for both men and women. The utility of these ratios is limited by the increasingly high cumulative mortality for all strata, approaching 100% in the older age groups on 20-year follow-up. Therefore, subsequent mortality analyses are based on hazard rates and multivariable models of time to death.

The mortality hazard rates and black/white ratios by cause of death and index of social status for men are displayed in table 4. The black/white ratios for CHD are less than one for both age and white social status comparison strata. For CVD and all causes at ages 45–64, the black/white mortality ratios are close to one with low social status whites as the referent and to 1.5 with high social status white males as the referent group. For each age and cause-of-death group, the black/white mortality hazard rate ratio is higher for high social status than for low social status men; i.e. the mortality rate is higher among low social status than high social status white men. The black/white mortality rate ratio is lower in men over age 65 than in men ages 40–64 for each cause-of-death/social status comparison stratum.

The mortality hazard rates and black/white ratios by cause of death and index of social status for women are presented in table 5. For each age stratum, cause of death (CHD, all CVD and all causes) and each white social status comparison stratum, the age-adjusted mortality (hazard) rate is higher for black than white women with the exception of CHD deaths for black middle-aged compared with white middle-aged lower social status women. For each age group and cause of death (except for deaths attributed to CHD among ≥65-year-old women), the hazard rate ratio is higher comparing black with high social status than low social status women; i.e. the mortality rate is lower in high social status than low social status women; i.e. the mortality rate is lower in high social status than low social status white women. The black/white ratio is lower among women aged ≥65 than among women aged 40–64 years for each social status/cause-of-death stratum except for deaths attributed to CHD among low social status women.

Kaplan-Meier survival curves for middle-aged and older males, quantitatively describing the experience over time prior to death from CHD, CVD and all causes are displayed in figure 3. For both middle-aged and older men, there is a longer period of survival free

Table 4. Mortality rate (±SEM) and black/white ratio by cause of death and index of social status: men

	46–64			≥65		
	black	LSS	HSS	black	LSS	HSS
CHD						
Dead	33	41	36	16	17	19
Age-adjusted rate	7.6±1.3	10.0±1.6	8.8±1.5	23.1±5.7	42.0±10.0	31.5±7.1
B/W ratio		0.76	0.86		0.55	0.73
CVD						
Dead	75	74	52	36	28	39
Age-adjusted rate	17.7±2.1	17.7±2.1	12.7±1.8	52.0±8.4	69.5±12.6	66.1±10.2
B/W ratio		1.0	1.4		0.75	0.79
All causes						
Dead	137	136	89	63	50	56
Age-adjusted rate	31.8±2.7	32.3±2.7	21.9±2.3	89.5±10.7	123.0±16.2	95.4±12.0
B/W ratio		0.98	1.5		0.72	0.94

LSS = Low social status; HSS = high social status; B/W = black/white.

Table 5. Mortality rate (±SEM) and black/white ratio by cause of death and index of social status: women

	46–64			≥65		
	black	LSS	HSS	black	LSS	HSS
CHD						
Dead	24	24	17	26	18	26
Age-adjusted rate	4.0±0.8	5.3±1.1	3.2±0.8	27.2±5.3	22.7±5.3	23.2±4.6
B/W ratio		0.75	1.3		1.2	1.2
CVD						
Dead	68	38	27	52	32	39
Age-adjusted rate	11.4±1.4	8.4±1.3	4.8±1.0	54.0±7.1	41.0±7.2	35.1±5.6
B/W ratio		1.4	2.4		1.3	1.5
All causes						
Dead	120	67	58	70	49	58
Age-adjusted rate	20.1±1.8	14.6±1.8	10.8±1.4	73.6±8.3	62.5±8.6	52.1±6.9
B/W ratio		1.4	1.9		1.2	1.4

of CHD as a cause of death for black than white higher or lower social status men during most of the follow-up period. This survival advantage is eliminated for the initially middle-aged cohort at 19 years follow-up and for the initially older cohort at 16 years follow-up. A selective survival advantage for black middle-aged men is not present for deaths attributed either to all CVD or all causes. Their experience is virtually identical to that

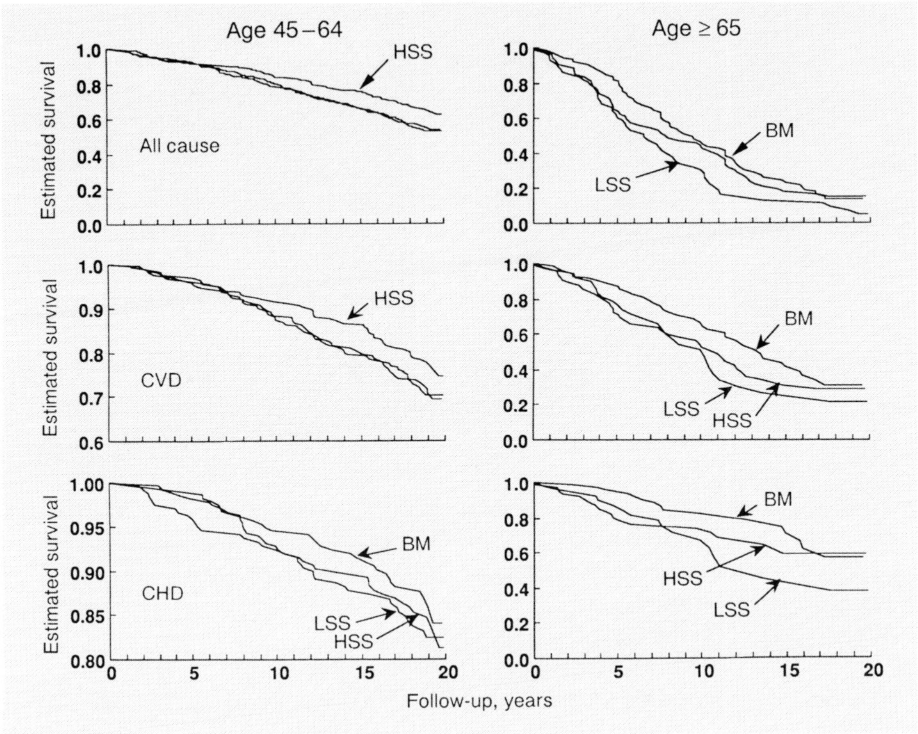

Fig. 3. Kaplan-Meier survival curves for males in two age groups by index of social status and cause of death. For abbreviations, see legend to figure 1.

of low social status white men. In contrast, high social status white men exhibit greater survival beginning 5 years after initiation of follow-up and persistent through 20 years. The pattern of survival by race/SES for both all CVD and all causes is different for men aged 65 or older. Among these older men, survival is consistently better for older black men until 17 years of follow-up. Among older white men, survival for CHD, all CVD and all causes is generally better for higher than for lower social status men.

Kaplan-Meier survival curves for middle-aged and older females, quantitatively describing the experience over time prior to death from CHD, CVD and all causes, are displayed in figure 4. In contrast to the apparent selective survival advantage of black men for CHD, no major clear differences in patterns of survival from CHD deaths appear among women. However, there is marked, consistently lower survival for black females for CVD and all-cause mortality for both middle-aged and older women. No consistent pattern of survival differentiates low from high social status white females across cause-of-death and age strata.

Risk Functions for Mortality Associated with Serum Cholesterol, Blood Pressure and Smoking

The multivariable association of risk factors for CHD mortality assessed by proportional hazards models of time to death among men and women stratified by age groups is set out in table 6. For both middle-aged and older

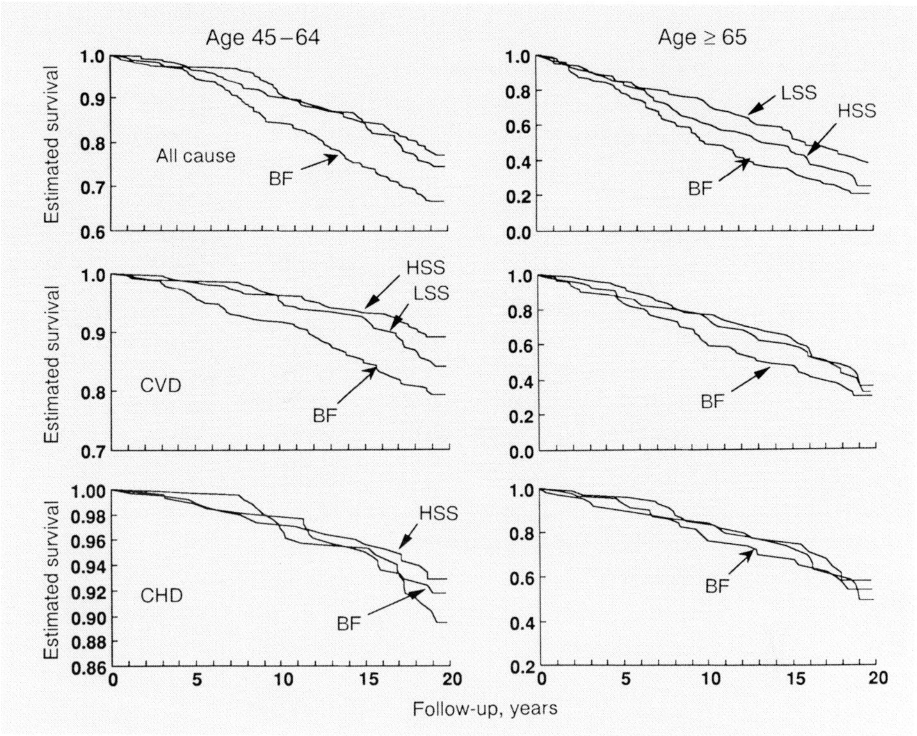

Fig. 4. Kaplan-Meier survival curves for females in two age groups by index of social status and cause of death. For abbreviations, see legend to figure 1.

Table 6. Multivariate association of risk factors with CHD mortality among middle-aged and older men and women in Evans County

Risk factors	Men				Women			
	40–64 (n = 835)		≥65 (n = 191)		40–64 (n = 892)		≥65 (n = 229)	
	B	SE	B	SE	B	SE	B	SE
Age	0.063[a]	0.016	0.009	0.048	0.036	0.021	0.095[b]	0.038
Low SES white[1]	0.068	0.240	0.118	0.364	0.238	0.325	0.251	0.342
Black[1]	–0.428[c]	0.264	–1.133[a]	0.390	–0.225	0.357	0.001	0.326
Prevalence of CHD	0.585	0.387	0.381	0.498	1.213[a]	0.340	0.478	0.629
SBP	0.015[a]	0.003	0.015[a]	0.005	0.015[a]	0.004	0.017[a]	0.004
Current smoking	0.512[b]	0.213	0.847[a]	0.316	–	–	–	–
Cholesterol	0.002	0.003	0.010[a]	0.004	0.008[a]	0.002	0.003	0.003
BMI	0.009	0.024	–0.050	0.040	0.047[b]	0.018	0.002	0.024
Major ECG	0.643[b]	0.311	1.648	0.424	0.557	0.353	0.850[a]	0.319
Minor ECG	0.201	0.231	0.643	0.416	–0.216	0.357	0.269	0.345

[1] Referent group is high SES white men and women, respectively.
[a] $p \leq 0.01$, [b] $p < 0.05$, [c] $p < 0.10$.

Table 7. Multivariate association of risk factors with CVD mortality among middle-aged and older men and women in Evans County

Risk factor	Men				Women			
	40–64 (n = 835)		≥65 (n = 191)		40–64 (n = 892)		≥65 (n = 229)	
	B	SE	B	SE	B	SE	B	SE
Age	0.074[a]	0.012	0.052[c]	0.032	0.057[a]	0.015	0.107[a]	0.029
Low SES white[1]	0.280	0.190	–0.045	0.266	0.303	0.256	0.245	0.262
Black[1]	0.042	0.198	–0.859[a]	0.261	0.349	0.252	0.226	0.248
Prevalence of CHD	0.687[c]	0.285	0.048[a]	0.355	0.629[c]	0.348	0.358	0.482
SBP	0.015[a]	0.002	0.014[a]	0.003	0.016[a]	0.002	0.016[a]	0.003
Current smoking	0.588[a]	0.159	0.397[c]	0.226	–	–	–	–
Cholesterol	0.002	0.002	0.009[a]	0.003	0.006[a]	0.002	–0.001	0.002
BMI	–0.006	0.018	–0.024	0.027	0.019	0.013	–0.000	0.018
Major ECG	0.818[a]	0.212	0.995[a]	0.283	0.544[b]	0.234	0.488[b]	0.243
Minor ECG	–0.020	0.178	0.161	0.279	0.029	0.228	0.116	0.251

1 Referent group is high SES white men and women, respectively.
[a] $p \leq 0.01$, [b] $p < 0.05$, [c] p 0.10.

Table 8. Multivariate association of risk factors with all-cause mortality among middle-aged and older men and women in Evans County

Risk factor	Men				Women			
	40–64 (n = 835)		≥65 (n = 191)		40–64 (n = 892)		≥65 (n = 229)	
	B	SE	B	SE	B	SE	B	SE
Age	0.065[a]	0.009	0.038	0.025	0.048[a]	0.011	0.133[a]	0.024
Low SES white[1]	0.324[b]	0.143	0.171	0.209	0.162	0.182	0.292	0.213
Black[1]	0.106	0.149	–0.403[b]	0.203	0.281	0.177	0.135	0.207
Prevalence of CHD	0.482[b]	0.226	0.339	0.272	0.630[b]	0.264	0.690[b]	0.350
SBP	0.012[2a]	0.002	0.007[a]	0.003	0.012[a]	0.002	0.016[a]	0.002
Current smoking	0.681[a]	0.121	0.189	0.173	–	–	–	–
Cholesterol	0.004[a]	0.001	0.003	0.002	0.004[a]	0.001	–0.002	0.002
BMI	–0.289[b,*]	0.105	–0.024	0.020	–0.001	0.011	–0.001	0.015
BMI^2	0.005[b,*]	0.002	–	–	–	–	–	–
Major ECG	0.661[a]	0.162	0.855[a]	0.222	0.567[a]	0.174	0.535[a]	0.200
Minor ECG	–0.134	0.132	0.232	0.211	0.096	0.167	0.092	0.212

1 Referent group is high SES white men and women, respectively.
[a] $p \leq 0.01$, [b] $p < 0.05$. * = Significance level is based on both the linear and the quadratic component.

men the hazard for CHD mortality is lower comparing blacks with white high social status men, adjusting for the major risk factors, age, BMI, existence of prevalent CHD and the presence of ECG abnormalities at baseline. Although the coefficient for the proportional mortality hazard ratio of low compared with high social status white men is positive, the coefficient is not statistically significant in either age group. In models for both age groups, after controlling for race/SES, age, BMI, prevalence of CHD and ECG abnormalities, the risk factors of blood pressure and current smoking are statistically significant in middle-aged men; additionally, cholesterol is statistically significant in older men.

The multivariable association of risk factors for CHD mortality among women stratified by age groups are considered next. In contrast to the findings for men, the hazard ratio for CHD is not statistically significant comparing black with high social status white women in either age group (although the coefficient was negative for middle-aged women). Although the coefficient contrasting the mortality hazard rate in low with high social status white women is positive, the coefficient is not statistically significant in either age group. Within these models, controlling for race/SES, age, BMI, prevalence of CHD and ECG abnormalities, systolic blood pressure is statistically significant for both middle-aged and older women; cholesterol is statistically significant, however only for middle-aged women. Smoking was deleted from the models for women due to the very low prevalence of smoking among women in this population.

Multivariable proportional hazard models for CVD mortality among middle-aged and older men and women are set out in table 7. The CVD mortality rate for black men is higher comparing black with high SES white males in the middle-aged group, however, this is not statistically significant; in contrast, the hazard rate is statistically significantly lower for blacks for CVD mortality among older men, controlling for the other variables in the model. Systolic blood pressure and smoking are statistically significant predictors in both age groups; additionally, cholesterol is a statistically significant predictor in the older men.

In contrast to the findings for men, the hazard rate is greater for black than high SES white women in each age group; however, neither is statistically significant. Systolic blood pressure is a significant predictor in both age groups, and cholesterol is a significant predictor in middle-aged women. Low SES white women have higher CVD mortality than high SES white women, however these findings were not statistically significant.

Multivariable proportional hazard models for all-cause mortality among middle-aged and older men and women are set out in table 8. Middle-aged black men have higher all-cause mortality than high SES white men although this difference is not statistically significant; in contrast, mortality of older black men is statistically significantly lower than that of high SES white men. Systolic blood pressure is a significant predictor of all-cause mortality in both age groups; however, cholesterol and smoking status, while positive predictors in both age groups are statistically significant only for middle-aged men. Low SES white men have higher mortality rates than high SES white men in both age groups. The difference is statistically significant only in middle-aged men.

The mortality rate is higher for black than for high SES white women in both age groups, although neither difference is statistically significant. Systolic blood pressure is a significant predictor in each age group, and serum cholesterol is a significant predictor in middle-aged women. Low SES white women had higher mortality than high SES white women

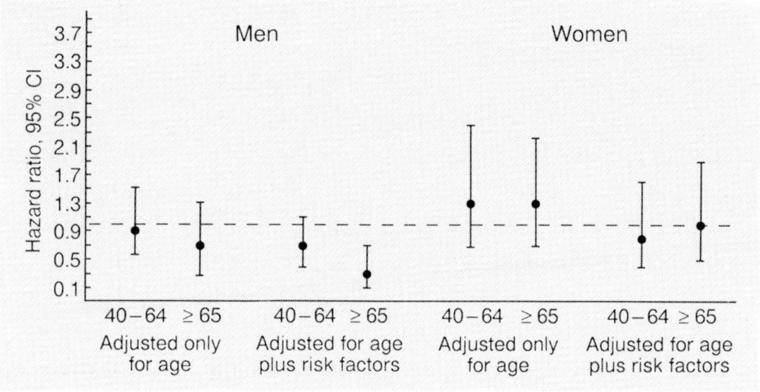

Fig. 5. Black/white CHD mortality hazard ratios. CI = Confidence interval. Black/white ratio = black/white high social status.

in both age groups, but neither difference is statistically significant.

In each of the cause-of-death-, gender- and age-specific models discussed above, either prevalence of CHD and/or the presence of major ECG abnormalities at baseline is a statistically significant predictor of death.

The coefficients derived from the multivariable proportional hazard analyses were converted to hazard ratios of CHD mortality for black compared with white high social status males and females controlling for risk factors; they are set out by age groups in figure 5. Among males, the relative protection from CHD mortality among blacks, i.e. the presence of hazard rate ratios less than one, is apparent. Of particular interest, the apparent protectiveness present when controlling for age increases in magnitude upon controlling for major risk factors. It is also noticeable upon controlling for age only or controlling for age plus risk factors that the hazard rate ratio is lower among older than middle-aged men. In contrast to the patterns of apparent relative protectiveness from CHD death among black men which increased at older ages, there is no similar evidence among women. Among women, the black/white hazard rate ratio associated with CHD is greater than the one among middle-aged and older women controlling for age only and increases slightly to one in older women controlling for age plus risk factors.

Discussion

The major question addressed by the analyses of this presentation was whether or not there was a difference in the 20-year CHD mortality experience of blacks and whites in a representative sample of the residents of Evans County in 1960. The results of the 20-year mortality follow-up experience confirmed the clinical observation of C.G.H.: the hazard of CHD mortality was less for black than for white males. This was not true for women: the CHD mortality hazard of black women was equal to or greater than that for white women. These results are based on causes of death as entered and coded on death certificates. They are subject to potential selection and information bias resulting from black/white differences in the seeking and obtaining of medical care and in possible differences in the certification of causes of death for blacks contrasted with whites. However, the mortality rate was lower in black than white men aged ≥65 not only for deaths attributed to CHD, but all CVD (and all causes). Moreover, similar findings regarding the relative CHD protectiveness of black males had previously been reported in Evans County in both prevalence [9] and 7-year incidence studies [10] in which objective, comparable criteria were applied to the diagnoses of CHD by members of a study panel.

The observed decrease in the black/white CHD mortality ratio with increasing age in the Evans County cohort is consistent with the black/white mortality cross-over observed in the US Vital Statistics since 1900. Analysis of the 20-year all-cause mortality experience of the Evans County cohort has previously been reported to show not just a decrease in the black/white ratio, but a statistically significant black/white mortality cross-over for men at age 73 and women at age 85 [11]. These patterns result from lower white mortality rates at younger ages and lower black rates at the oldest ages. These analyses suggest that in addition to the presence of possible CHD-protective factors in elderly black males, selective survival following periods of earlier mortality excesses may have been responsible for the cross-over.

The relative CHD protectiveness of black compared with white males in the setting of Evans County was not explained by differences in the standard risk factors, the prevalence of CHD at entry to the study, nor the prevalence of ECG abnormalities. In fact the relative protectiveness becomes more impressive in view of the markedly higher prevalence of elevated blood pressure and major ECG abnormalities in blacks than in whites. Each of these major determinants of CHD mortality in whites was previously found to be similarly predictive of CHD incidence in blacks [12]. They were found to be similarly predictive of CHD mortality in blacks and whites in the present study. Thus, the relative CHD protectiveness of black compared with white males must be due to strong CHD-protective factors not measured in this study. Data are not available directly to test for these potentially protective factors; however, other studies on subsamples of the cohort disclosed the following: major differences in lipoprotein fractions, i.e. higher serum HDL cholesterol, higher Apo-AI and lower levels of serum triglycerides in black than white males [5, 13]; higher levels of fibrinolytic activity [14] and higher levels of physical activity in black than white males [15], each of which is putatively antiatherosclerotic and/or antithrombotic, and which singly and/or in aggregate could explain the relative black male protectiveness regarding CHD. These findings are supportive of group differences reflecting responses to social, behavioral and life situational experiences. In addition, there were slight differences in the total fat and daily cholesterol intake [9]. Forty-two percent of calories were consumed as fat in black men compared with 45% among low social class white men, and daily cholesterol intake averages 650 mg in low social class whites and 527 mg in black men. Although the differences are small, the contrasts in dietary intake of fat and cholesterol between blacks and whites are similar to the direction of differences in serum cholesterol levels, i.e. lower in blacks than whites. It should be noted, however, that the major difference between the mortality experience of black and white men was in the comparison of black with high SES white men; there was little difference in the mortality experience of black compared with white low SES men.

It is important to emphasize that the evidence in Evans County of some protectiveness of black males against CHD does not indicate 'immunity'. The earlier literature on this subject led Gillum [16] to identify the 'myth' of black 'immunity' to CHD as harmful to the medical care and public health status of blacks. The findings presented in this study are based on the mortality experience of a predominantly rural, southern community in the 1960s and 1970s. The findings are not necessarily generalizable to other places nor necessarily to Evans County in more recent time periods. For example, studies of the white population have demonstrated a reversal of the social class relationship to CHD [17]. The data are not available to test for similar effects in blacks. Further, in consideration of the relative protectiveness of the black male against CHD mortality in Evans County, it is important to note that despite this relative protectiveness, CHD emerged as the leading single cause of death of black males.

The 20-year follow-up experience of the Evans County cohort confirmed the importance of the major risk factors of serum cholesterol, smoking and blood pressure for middle-aged and older, black and white men and women for CHD, CVD and all-cause mortality. Black men were less likely to die of CHD than white men, and this relative protectiveness increased upon adjusting for risk factors and also increased with age. The relative protectiveness of black men for CHD mortality was also reflected in lower black than white CVD and all-cause mortality in men 65 years of age and older.

Acknowledgements

This work was supported in part by the Merit Award NIH 2-R01-ML03341 and by the NIH R01 HL31397. We gratefully acknowledge the secretarial assistance of Lesa McPherson.

References

1 Hames CG: Evans County Cardiovascular and Cerebrovascular Epidemiologic Study: Introduction. Arch Intern Med 1971;128:883–886.
2 McDonough JR, Hames CG, Stubl SC, Garrison GE: Cardiovascular disease among negroes and whites in Evans County, Georgia. J Chronic Dis 1965;18:443–455.
3 Cassel J, Patrick R, Jenkins D: Epidemiological analysis of the health implications of culture change: A conceptual model. Ann NY Acad Sci 1960;84:938.
4 Keys A: Diet and the epidemiology of coronary heart disease. JAMA 1957;164:1912.
5 Tyroler HA, Hames CG, Krishan I, Heyden S, Cooper G, Cassel JC: Black-white differences in serum lipids and lipoproteins in Evans County. Prev Med 1975;4:541.
6 Schoenbach VJ, Kaplan BH, Fredman L, Kleinbaum DG: Social ties and mortality in Evans County, GA. Am J Epidemiol 1986;123:577–591.
7 McDonough JR, Hames CG: Influence of race, sex and occupation on seasonal changes in serum cholesterol. Am J Epidemiol 1967;83:356.
8 McGuire C, White GD: The measurement of social status. Research Paper in Human Development No. 3 (revised). Department of Educational Psychology, University of Texas, 1955.
9 Cassel JC: Review of 1960 through 1962 cardiovascular disease prevalence study. Arch Intern Med 1971;128:890–895.
10 Cassel JC, Heyden S, Bartel AG, Kaplan BH, Tyroler HA, Cornoni JC, Hames CG: Incidence of coronary heart disease by ethnic group, social class, and sex. Arch Intern Med 1971;128:901–906.
11 Wing S, Manton KG, Stallard E, Hames CG, Tyroler HA: The black/white mortality crossover: Investigation in a community-based study. J Gerontol 1985;40:78.
12 Kleinbaum DG, Kupper LL, Cassel JC, Tyroler HA: Multivariate analysis of risk of coronary heart disease in Evans County, GA. Arch Intern Med 1971;128:943–948.
13 Tyroler HA, Heiss G, Schonfeld G, Cooper G, Heyden S, Hames CG: Apolipoprotein A-I, A-II and C-II in black and white residents of Evans County. Circulation 1980;62:249.
14 Dischinger P, Tyroler HA, McDonagh R, Hames CG: Blood fibrinolytic activity, social class and habitual physical activity – I. A study of black and white men in Evans County, Georgia. J Chronic Dis 1980;33:283.
15 Skinner JS, Benson H, McDonough JR, Hames CG: Social status, physical activity, and coronary proneness. J Chronic Dis 1966;19:773–783.
16 Gillum RF: Coronary heart disease in black populations. I. Mortality and morbidity. Am Heart J 1982;104:839.
17 Morgenstern H: The changing association between social status and coronary heart disease in a rural population. Soc Sci Med 1980;14A:191–201.

Cardiology 1993;82:137–152

William B. Kannel[a]
Martin Larson[b]

[a] Section of Preventive Medicine and Epidemiology, Boston University School of Medicine, Evans Memorial Department of Clinical Research, University Hospital at Boston University Medical Center, Boston, Mass., USA;
[b] Framingham Study and National Heart, Lung, and Blood Institutes, Framingham, Mass., USA

Long-Term Epidemiologic Prediction of Coronary Disease

The Framingham Experience

Abstract

Atherosclerotic cardiovascular disease is a complex problem involving lipid deposition, pressure, rheologic forces, carbohydrate tolerance and thrombogenesis. The major contributors identified through epidemiologic research include atherogenic personal attributes, living habits which promote them, signs of a compromised coronary circulation and host susceptibility to these risk factors. Of the atherogenic risk attributes, such as blood lipids, blood pressure, glucose tolerance and fibrinogen, each independently contributes to risk, and the risk associated with any one is compounded by the presence of the others. The risk associated with hypertension, hyperlipidemia or diabetes varies widely depending on the level of associated risk factors. Also, at a given level of total cholesterol, risk is greatly affected by the total/HDL cholesterol ratio, which provides a practical means for assessing the two-way traffic of cholesterol. In addition, living habits, such as cigarette smoking or lack of exercise, can independently affect the risk associated with any of the atherogenic traits. These living habits, obesity and diet can also affect the level of atherogenic risk factors and must be taken into account in assessing risk and implementing preventive measures. Finally, preclinical indicators of silent myocardial ischemia greatly augment the risk associated with a poor cardiovascular risk profile. Hence, ECG left ventricular hypertrophy, blocked intraventricular conduction, repolarization abnormalities and abnormal response to exercise on monitoring must be taken into consideration. Optimal risk predictions require a quantitative synthesis of risk factors into a composite estimate. Handbooks, hand calculators and PC software have been devised for office use based on multiple logistic risk formulations. These have been shown to accurately predict disease risk in a variety of American population samples, in elderly as well as young coronary candidates. Preventive management as well as risk estimation should be multifactorial if optimal results are to be achieved. Preventive strategies should include public health measures to alter the ecology so as to shift the distribution of risk factors to a more favorable level, health education to enable people to protect their own health and preventive medicine for high-risk candidates. Greater skill must be developed to carry out such interventions. In selecting drugs to correct hypertension, diabetes and lipid disorders, it is important to choose agents which do not adversely affect the composite risk profile.

Key Words

Framingham Study
Preventive management
Risk factors, atherogenic
Predictability

Supported by the National Heart, Lung, and Blood Institute grants NIH-N01-HV-92922 and NIH-N01-HV-52971, the Charles A. Dana Foundation, Merck, Sharp and Dohme, Pfizer, and ICI.

William B. Kannel
Framingham Study
5 Thurber St.
Framingham, MA 01701 (USA)

Table 1. Classes of cardiovascular risk factors

Living habits	
Overeating; lack of exercise; cigarette smoking, Type A behavior?	
Atherogenic personal attributes	
High blood pressure; hyperglycemia; dyslipidemia; elevated fibrinogen	
Indicators of compromised circulation	
ECG abnormalities at rest; on exercise; on ambulatory monitoring; vascular bruits; echocardiographic abnormalities; myocardial perfusion deficits	
Host susceptibility	
Inborn errors of metabolism; family history of premature cardiovascular disease	

Introduction

Four decades of prospective longitudinal investigation of the evolution of coronary heart disease (CHD) at Framingham and elsewhere have provided insights into the factors which predispose to it [1]. Because the associations found are observational and not based on controlled experiments, the predisposing conditions have been termed 'risk factors'. Epidemiologic research has delineated multiple factors associated with the subsequent development of disease, and no single factor has been found to be essential or sufficient in the evolution of the disease. This is quite consistent with biologic research into the evolution of the atherosclerotic plaque which incriminates lipids, rheologic forces, platelet aggregation, growth factors, blood pressure, insulin resistance, smooth muscle proliferation, fibrosis, cell membrane receptors, inflammation and macrophages among others. Hence, an influence of multiple factors seems plausible.

Epidemiologic data from the Framingham Study suggest five major classes of risk factors which promote the disease: (1) faulty lifestyles, (2) atherogenic biologic attributes, (3) signs of active lesion degeneration, (4) indications of a compromised circulation and (5) host susceptibility (table 1). Assessment of risk and optimal treatment must encompass these multivariate influences. Also, these multiple factors provide diverse opportunities for pharmaceutical and hygienic interventions to interrupt the chain of events leading to the disease.

Need for a Preventive Approach

The longitudinal population approach to investigation of the evolution of coronary disease has made it apparent that coronary disease is extremely common and highly lethal. One in every six coronary attacks is found to present with sudden death as the first, last and only symptom. Most coronary deaths occur outside the hospital because half of them occur suddenly without warning. One in three myocardial infarction goes unrecognized because of lack of symptoms or symptoms so atypical as to be ascribed to other causes.

With this kind of disease, awaiting the appearance of symptoms is a form of brinkmanship that cannot be condoned. A coronary attack must come to be regarded as a medical failure rather than the first indication for treatment. Rather, we must identify coronary candidates well in advance of symptoms, and detect and correct modifiable predisposing factors. Fortunately, this now seems feasible since epidemiologic investigations at the Framingham Study and elsewhere have identified a number of predisposing factors which can be used to target susceptible persons for preventive measures long in advance of symptoms.

Table 2. Net impact of specified risk factors on cardiovascular disease outcomes: 30-year follow-up, Framingham Study

Risk factors	Increment per standard deviation increase, %							
	coronary disease		stroke		peripheral vascular disease		congestive heart failure	
	men	women	men	women	men	women	men	women
Blood pressure								
Systolic	33***	42***	63***	49***	33***	40***	28**	37**
Diastolic	33***	34***	69***	50***	17[a]	24*	12[a]	34**
Serum cholesterol	34***	30***	–1[a]	–4[a]	29**	25**	25*	–4[a]
Hematocrit	0.6[a]	8	39***	15[a]	26**	–1[a]	9[a]	36**
Glucose	4[a]	18***	4[a]	14*	22***	29***	29***	25***
Met. relative weight	16***	15**	6[a]	8[a]	30**	1[a]	–	38***
Vital capacity	–3[a]	–26***	–7[a]	–28*	–19[a]	–3[a]	–48	100***
Heart rate	5[a]	–3[a]	14[a]	7[a]	12[a]	5[a]	42***	13[a]
Cigarettes	34***	15**	38***	10[a]	61***	53***	30**	15[a]

Multivariate risk: taking systolic blood pressure, cholesterol, glucose, cigarettes and LVH into account.
* $p < 0.05$, ** $p < 0.01$, *** $p < 0.001$.
[a] Nonsignificant.

Major Identified Risk Factors

Epidemiologic research at the Framingham Study has identified a number of metabolic and physiologic traits which promote accelerated atherogenesis, eventuating in major clinical atherosclerotic events [1]. Some nontrivial differences exist in their impact on the occurrence of the various cardiovascular outcomes (table 2). All the major identified cardiovascular risk factors, including elevated cholesterol, blood pressure, blood glucose, fibrinogen and cigarette smoking, predict CHD (table 2). For stroke, hypertension predominates, atrial fibrillation and cardiac failure make a major impact, but lipids have little influence (table 2). For peripheral arterial disease, cigarettes, blood pressure, cholesterol and glucose tolerance are most influential (table 2). For cardiac failure, blood pressure, left ventricular hypertrophy (LVH), low vital capacity, rapid heart rate and diabetes are paramount.

These common risk factors operate in both sexes, at all ages, but with different strengths. Diabetes and a low HDL cholesterol eliminate the female advantage over men (table 3, fig. 1). Cigarette smoking has a greater influence in men and is noncumulative and reversible on quitting (table 2). Fibrinogen is a major risk factor for coronary disease, stroke and peripheral arterial diseases in both sexes (table 4). Some risk factors, such as blood lipids, impaired glucose tolerance and fibrinogen, diminish in impact with advancing age, but continue to have relevance in the elderly because the lower risk ratios are offset by a higher incidence of CHD, a greater prevalence of the risk factors and a greater excess risk.

Obesity and weight gain promote all the major atherogenic traits, and physical indolence promotes risk factors and coronary dis-

Table 3. CHD risk in diabetics by age and sex: 36-year follow-up, Framingham study

	Age: 35–64 years		Age: 65–94 years	
	men	women	men	women
Biennial age-adjusted rate/1,000	38.8	41.6	79.4	62.2
Age-adjusted risk ratio	1.5	3.7	1.6	2.1
Excess risk	12.2***	30.2****	28**	32****

** $p < 0.01$, *** $p < 0.001$, **** $p < 0.0001$.

Table 4. Risks of specified cardiovascular (CV) events by tertile of fibrinogen: 18-year follow-up, Framingham study (subjects aged 47–79 years)

CV disease outcome	Absolute risk/1,000			
	risk (T_3)		ratio T_3/T_1	
	men	women	men	women
CHD	430	296	1.6**	2.6***
Stroke	187	224	1.9[a]	2.0[a]
Peripheral arterial disease	113	68	1.6[a]	1.7*
CV	611	514	1.5***	2.1***

T_3 = Third tertile of fibrinogen; T_1 = first tertile of fibrinogen.
[a] Nonsignificant.
* $p < 0.05$, ** $p < 0.01$, *** $p < 0.001$.

ease at all ages. Systolic pressure and isolated systolic hypertenison are major risk factors. The total/HDL cholesterol ratio has been found to provide the best and most convenient lipid risk profile.

While all these individual risk factors influence the risk of developing CHD, they seldom occur in isolation and, when they cluster, each augments the risk associated with the others, making it necessary to deal with any risk factors as an ingredient of a cardiovascular risk profile.

Predictability of CHD

Over the past 4 decades, there has been an inpressive increase in our ability to predict CHD. To the core triad of CHD risk factors (blood pressure, cholesterol and cigarette smoking), there have been added LVH, glucose intolerance, HDL cholesterol, fibrinogen and leukocyte count. Prediction can be evaluated as a problem of classification, hazard estimation or dose-response. Perfect discrimination occurs when each and every person in

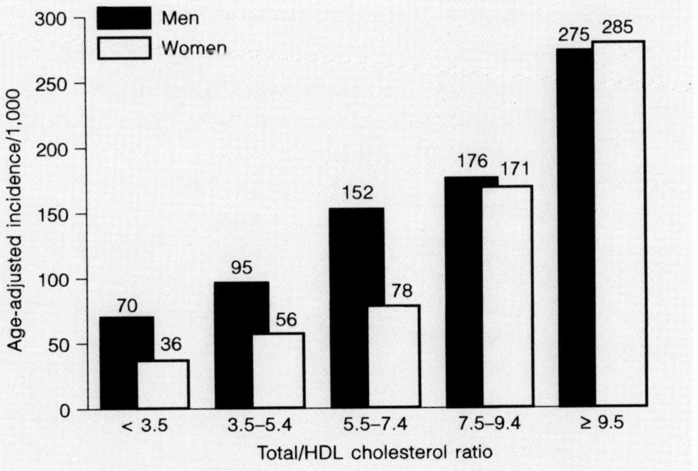

Fig. 1. Risk of CHD by total/HDL cholesterol ratio in 50- to 90-year-old participants: Framingham Study; 26-year follow-up.

a sample is either correctly identified as a case or not. This is unrealistic since there is some risk even in low-risk persons, and, if the prevalence of this class of persons is large enough, some cases must be expected. Nor can there be absolute certainty that any person will develop CHD, however high the apparent risk. A hazard approach assumes that every person of a given age who is free of CHD will develop CHD sooner or later unless he dies of some other cause first. Hence, we are concerned with the distribution of time to CHD conditional on risk factor makeup. Predictive power of risk factors is expressed in terms of the variance of this distribution.

A related approach is to consider the probability of CHD in a limited time span conditional on a specified set of risk factors in a dose-response relationship. As in the hazard formulation, nobody is certain to either avoid or develop CHD. Rather, the risk is considered to increase with the level of risk factors. The agreement of the actual and expected number of cases in each decile of multivariate risk can be examined to assess the accuracy of the prediction (i.e. goodness of fit). It is also desirable that the risk slope be steep; only a few cases should fall into the lowest risk decile, while a very large percent of cases should be in the highest risk decile (a sort of classification test). However, the group with intermediate risk ordinarily accounts for a large proportion of the population and of the CHD incidence. It is apparent that what is required is not a perfect risk classification, but one that gives as steep a risk gradient as possible. This can be handily evaluated by examining the risk ratio for upper and lower deciles of multivariate risk function.

Epidemiologic data from Framingham and elsewhere have established that CHD is a multifactorial disease. The personal characteristics that have been shown to be related to CHD risk include: blood lipids, blood pressure, cigarette smoking, glucose intolerance, LVH, fibrinogen and leukocyte (among others). It is important to recognize that none of these risk factors is strictly determinative and that in every instance the risk associated with any factor has been found to vary according to

Table 5. Impact of risk factors for initial CHD events by age and sex: 36-year follow-up

Risk factor	Age-adjusted rate/1,000 with RF	Age-adjusted rate/1,000 without RF	Risk ratio	Excess risk	p value
Males, 35–64 years old					
SCL	34.0	18.1	1.88	15.9	<0.0001
HBP	45.4	22.7	2.00	22.7	<0.0001
Diabetes	38.8	26.6	1.46	12.2	0.001
ECG-LVH	78.6	26.5	2.97	52.1	<0.0001
Smoking	32.6	21.2	1.54	11.4	<0.0001
Males, 65–94 years old					
SCL	58.5	49.5	1.18	9.0	0.24
HBP	72.6	44.9	1.62	27.7	<0.0001
Diabetes	79.4	51.1	1.55	28.3	0.002
ECG-LVH	137.9	51.9	2.66	86.0	<0.0001
Smoking	52.7	53.8	0.98	–	0.92
Females, 35–64 years old					
SCL	15.1	8.6	1.76	6.5	0.004
HBP	21.3	9.5	2.24	11.8	<0.0001
Diabetes	41.6	11.4	3.65	30.2	<0.0001
ECG-LVH	54.8	11.9	4.61	42.9	<0.0001
Smoking	12.9	12.1	1.07	0.8	0.62
Females, 65–94 years old					
SCL	39.2	19.8	1.98	19.4	0.0007
HBP	44.2	23.7	1.86	20.5	<0.0001
Diabetes	62.2	30.2	2.06	32.0	<0.0001
ECG-LVH	93.5	31.5	2.97	62.0	<0.0001
Smoking	37.6	32.1	1.17	5.5	0.28

SCL = Serum cholesterol > 240 vs. < 200; HBP = high blood pressure = ⩾ 160/95 mm Hg; RF = risk factor.

the associated constellation of other factors [1–3]. Multivariate assessment allows identification of persons with any one risk factor who are truly in jeopardy. Another reason for using multivariable risk assessments is that a constellation of risk factors provides substantially better prediction than one single risk factor.

Efficient methods for combining the predictive information from a number of risk factors conjointly are now available [2, 4–7]. By prediction we mean to estimate more precisely that persons with certain characteristics will develop CHD in some specified period of time. The risk factors involved are those identified as conferring a greater than average probability of developing CHD. Several risk factors must be considered conjointly if persons at very high or very low risk are to be identified.

The variables chosen by the Framingham Study to construct a coronary risk profile

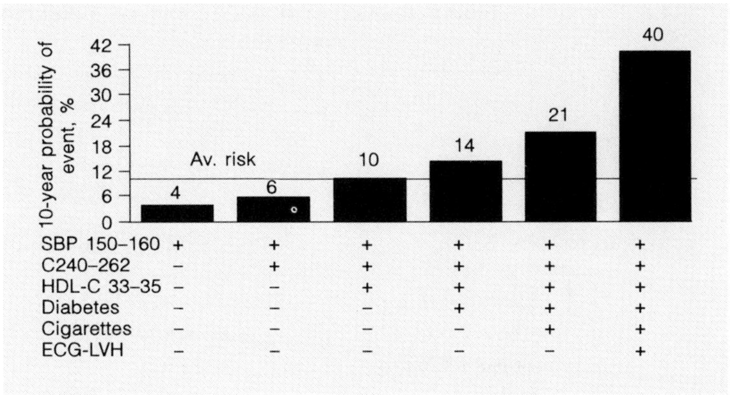

Fig. 2. Probability of CHD event in mild hypertension by intensity of associated risk factors in men aged 45 years: Framingham Study. SBP = Systolic blood pressure; C = cholesterol; Av. = average.

include serum total and HDL cholesterol, systolic blood pressure, glucose tolerance, cigarette smoking and ECG-LVH (fig. 2, table 5). These variables were chosen because each was found to be an independent predictor of CHD at all ages, in both sexes, and because they can be determined by ordinary office procedures and simple laboratory tests, without incurring great cost or inconvenience [1, 5, 8].

Serum total and HDL cholesterol were chosen because they can be obtained in a nonfasting state with reasonable accuracy. The serum total cholesterol reflects a two-way traffic of cholesterol being carried into the arterial intima in the LDL fraction and removed by the HDL cholesterol fraction (fig. 3). The total/HDL cholesterol ratio conveniently reflects this joint effect (fig. 1). The use of total cholesterol and HDL cholesterol provides the most efficient lipid prediction of CHD [1, 9]. This has been confirmed by the Physicians Health Study [8].

Systolic pressure was selected rather than diastolic pressure because it is more accurately measured, provides a wider range of values and is a more significant independent predictor. Adding diastolic as well as systolic pressure fails to improve the risk estimation, because the two components of blood pressure are too highly correlated (fig. 4). Elevated blood pressure is commonly associated with the other risk factors (table 6), and those in jeopardy have additional risk factors (fig. 2). There is no justification for the clinical preoccupation with the

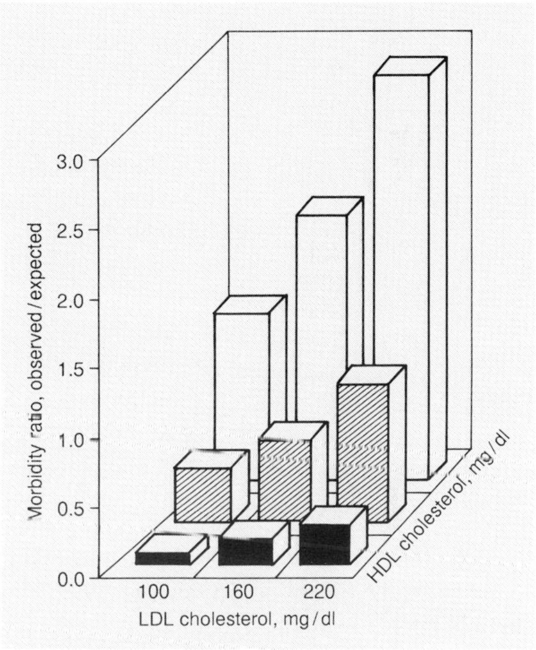

Fig. 3. CHD by HDL and LDL cholesterol in 50- to 70-year-old men: Framingham Study.

Table 6. Prevalence of cardiovascular (CV) risk factors and disease by hypertension status (in percent): Framingham Heart Study 1970–1982

	Age: 35–64					
	men			women		
	NORM	BORD	DEF	NORM	BORD	DEF
Person exams	2,581	826	650	3,298	686	443
CV risk factors						
Cholesterol ⩾ 240	20.8	25.0	31.4	25.7	32.8	35.3
HDL < 35	20.8	23.9	25.1	4.7	5.1	7.0
Diabetes	4.3	4.7	6.3	2.1	4.1	6.4
LVH (DEF)	0.4	0.9	3.2	0.2	0.4	1.8
Smokes cigarettes	43.8	36.9	38.2	39.0	34.9	34.4
Obese (MRW ⩾ 130)	25.8	36.6	50.8	20.7	34.5	44.2

Table 7. Risk of CHD by ECG and roentgenographic LVH: 32-year follow-up, Framingham Study (subjects 35–64 years)

ECG-LVH	Age-adjusted biennial rate/1,000			
	men: X-ray-LVH		women: X-ray-LVH	
	absent	present	absent	present
Absent	24	36	10	23
Present	85	100	45	62

diastolic blood pressure. Knowledge of the diastolic pressure offers no advantage when the systolic pressure is known (fig. 5).

ECG-LVH was included in the risk profile because it was found to be an ominous harbinger of atherosclerotic cardiovascular events at all ages, in both sexes, including CHD, its most common and most lethal sequela [10]. The ECG version of LVH is the most convenient and cost-effective, even though it is a less sensitive and specific indicator of anatomical hypertrophy than the echocardiographic or roentgenographic LVH entity. The ECG version contributes to risk independent of the anatomical version typified by chest X-ray (table 7). The risk of CHD associated with ECG-LVH risk varies widely depending on coexistent risk factors, and it independently contributes to the risk imposed by any one of them [10].

Glucose intolerance or diabetes is an important ingredient of the risk profile because it too makes an independent contribution to CHD risk. It is assigned a greater weight in women than men, and it cancels their CHD advantage over men (table 3). There is evi-

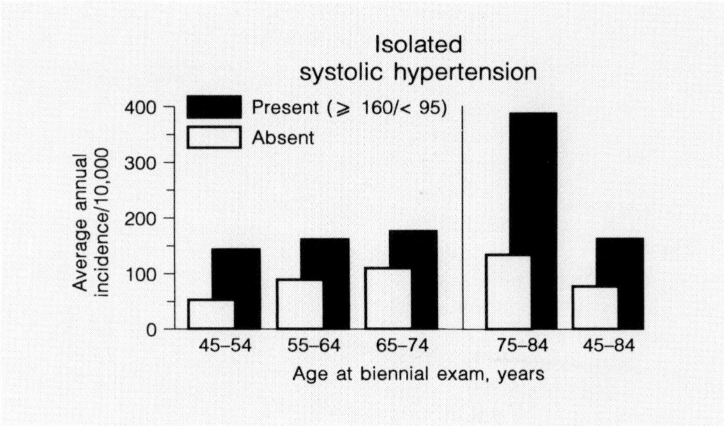

Fig. 4. Risk of myocardial infarction with isolated systolic hypertension (≥ 160/< 95 mm Hg). 24-Year follow-up in men 45–84 years old: Framingham Study. All ages: * $p < 0.01$ (age adjusted).

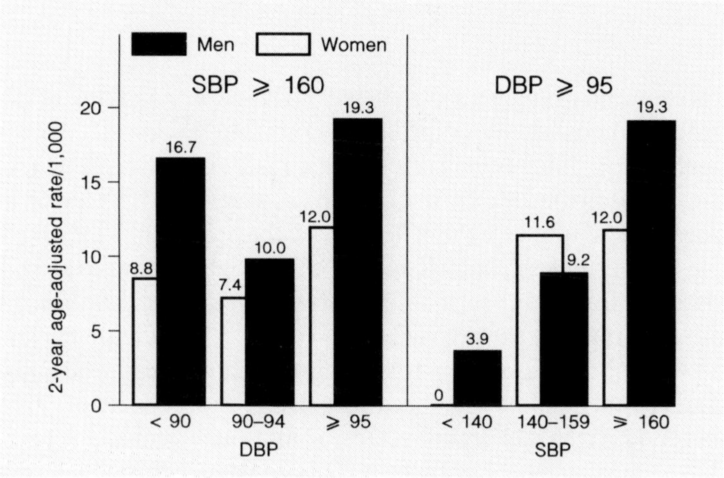

Fig. 5. Risk of stroke according to systolic (SBP) and diastolic blood pressure (DBP).

dence to suggest that insulin resistance may be fundamental to atherogenesis. It is metabolically linked to abdominal obesity, glucose intolerance, low HDL cholesterol, high triglyceride and hypertension [11]. The risk in diabetics varies widely depending on the extent to which they have these coexistent factors (fig. 6).

Cigarette smoking is an easily ascertained risk factor which is added because of its strong independent contribution to risk [12]. It adds to the risk associated with any of the other risk factors.

Another reason for the choice of these risk factors is that there are now available means for modifying them and, for most, evidence that CHD risk can be reduced by modifying one or more of them.

A number of factors which have been shown to contribute to CHD incidence have been excluded from the risk profile. Obesity is related to CHD incidence, particularly abdominal obesity which appears to contribute to risk at any body mass index. However, the effect is largely predicted by the risk factors promoted by obesity, such as glucose intoler-

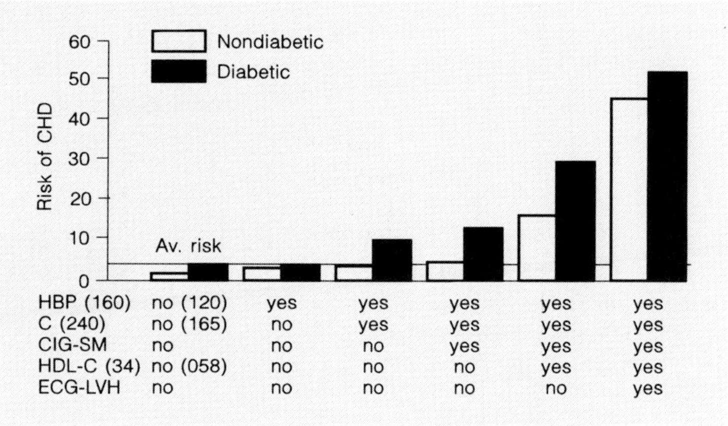

Fig. 6. Risk of CHD by diabetic status according to the level of other factors in 50-year-old women: Framingham Study. HBP = High blood pressure; C = cholesterol; CIG-SM = cigarette smoker; Av. = average.

ance, raised triglyceride, reduced HDL-cholesterol, increased blood pressure aand raised cholesterol. Since these factors are already in the profile, the addition of relative weight does little to improve the efficiency of the profile. Nevertheless, knowledge of the weight status is important in making recommendations to improve the profile.

A positive family history of premature CHD (under age 55 years) is also a strong risk factor [13]. However, its effect is also mediated by a tendency of such families to share elevated lipids, glucose intolerance, hypertension and obesity, which are already included in the profile. Taking these into account, the independent effect of a positive family history is only 30% in excess of those without such a history. Those with a positive family history should have a greater sense of urgency to detect and correct relevant risk factors.

Physical activity has an independent protective effect in the Framingham Study, as has been noted elsewhere. This is more evident in men than women, but seems protective at all ages (table 8). However, it is difficult to measure objectively using ordinary office procedures, so it has not been included in the risk profile. It is an important consideration in making recommendations to improve the risk profile since exercise can help in control of weight, glucose intolerance, HDL cholesterol, triglyceride and blood pressure.

Fibrinogen and leukocyte count are two candidates for inclusion in the risk profile in the future. Each is a strong independent risk factor for CHD (tables 4, 9). They may indicate the thrombogenic and inflammatory aspect of atherogenesis. Within the normal range of each factor, CHD risk increases progressively with the level of the factor. Active lesions undergoing fissuring, subintimal hemorrhage or inflammatory lipid infiltration may be indicated by the level of these two variables.

Multivariate Risk

The Framingham Study has repeatedly refined multivariate risk equations for use by clinicans and public health officials to predict the development of CHD in persons free of the condition [2–4]. In addition to age and sex, the risk factors have included systolic blood pressure, serum total cholesterol, cigarette smoking, glucose intolerance and left

Table 8. CHD mortality in men by physical activity index: 24-year follow-up, Framingham Study

Physical activity index[a]	24-year cumulative rate/1,000, years				
	45–49	50–54	55–59	60–64	45–64[b,c]
<29	250	147	255	404	255
29–34	114	177	264	198	184
>34	147	124	200	141	152

a Physical activity index at examination No. 4; population at risk = all persons alive.
b Age adjusted.
c $p < 0.01$ for trend of CHD mortality in relation to physical activity.

ventricular hypertrophy. Handbooks containing CHD risk tables based on Framingham Study multivariate equations were published by the American Heart Association and the National Heart, Lung, and Blood Institute in 1973 [5, 6]. A simpler approximation of the equations was again produced in 1982 [7]. The latest version has added HDL cholesterol [14]. The database used is larger and more recent and provides more data for persons over age 60 years. To provide current estimates of CHD risk over a large age range, data from the original cohort were combined with data from the second-generation study underway, for which 8 years of follow-up were available. Persons 30–74 years of age were included. The derivation and uses of this updated coronary risk profiles have been published recently [14].

The CHD risk equations, and the accompanying point score prediction probability algorithm, allow estimation of coronary candidates 5- and 10-year risks of CHD. These calculations are not appropriate for persons who already have CHD (table 10).

Generalization of the Framingham risk profile to the population at large must be done with caution. However, it has been repeatedly

Table 9. Cardiovascular disease incidence by white blood count – Framingham offspring. 12-year follow-up (subjects 30–59 years of age),

Tertile of white blood count	12-Year age-adjusted rate/1,000	
	men***	women*
1 (2,600–5,400)	52	38
2 (5,400–6,900)	122	43
3 (6,900–14,200)	169	71

Group trends statistically significantly:
* $p < 0.05$, *** $p < 0.001$.

shown that the Framingham risk model is effective in predicting CHD in other large population samples in the USA [15–18]. However, the equations may not be applicable in populations with very low CHD incidence rates [17]. They also may be inappropriate for persons with very high risks factors, such as would occur in malignant hypertension, severe diabetes, familial hypercholesterolemia or other conditions which place persons in the extreme percentiles of the risk distributions [19]. However, the Framingham

Table 10. CHD risk factor prediction chart

a Find points for each risk factor

Age (if female)		Age (if male)		HDL-cholesterol		Total-cholesterol		Systolic blood pressure		Other	Points
age	points	age	points	HDL-C	points	total-C	points	SBP	points		
30	–12	30	–2	25–26	7	139–151	–3	98–104	–2	Cigarettes	4
31	–11	31	–1	27–29	6	152–166	–2	105–112	–1	Diabetic-male	3
32	–9	32–33	0	30–32	5	167–182	–1	113–120	0	Diabetic-female	6
33	–8	34	1	33–35	4	183–199	0	121–129	1	ECG-LVH	9
34	–6	35–36	2	36–38	3	200–219	1	130–139	2		
35	–5	37–38	3	39–42	2	22–239	2	140–149	3	0 points for each NO	
36	–4	39	4	43–46	1	240–262	3	150–160	4		
37	–3	40–41	5	40–50	0	263–288	4	161–172	5		
38	–2	42–43	6	51–55	–1	289–315	5	173–185	6		
39	–1	44–45	7	56–60	–2	316–330	6				
40	0	46–47	8	61–66	–3						
41	1	48–49	9	67–73	–4						
42–43	2	50–51	10	74–80	–5						
44	3	52–54	11	81–87	–6						
45–46	4	55–56	12	88–96	–7						
47–48	5	57–59	13								
49–50	6	60–61	14								
51–52	7	62–64	15								
53–55	8	65–67	16								
56–60	9	68–70	17								
61–67	10	71–73	18								
68–74	11	74	19								

b Sum points for all risk factors

______	+	______	+	______	+	______	+	______	+	______	+	______	=	______
Age		HDL-C		Total-C		SBP		Smoker		Diabetes		ECG-LVH		Point total

Note: Minus points subtract from total.

risk profile can distinguish high- from low-risk persons based on their relative risk in low- or high-risk populations.

Appropriate use of the CHD risk profiles can be helpful in selecting patients at risk of CHD for preventive maintenance without needlessly alarming or falsely reassuring persons with some particular risk factor. It is also useful in monitoring progress of patients prescribed preventive measures. It can also be helpful in motivating patients to start or continue such measures. Clinicians can also use the profile to estimate the possible benefit to be gained by contemplated interventions.

Table 10 (continued)

c Look up risk corresponding to point total

Probability		
points	5 years, %	10 years, %
≤1	<1	<2
2	1	2
3	1	2
4	1	2
5	1	3
6	1	3
7	1	4
8	2	4
9	2	5
10	2	6
11	3	6
12	3	7
13	3	8
14	4	9
15	5	10
16	5	12
17	6	13
18	7	14
19	8	16
20	8	18
21	9	19
22	11	21
23	12	23
24	13	25
25	14	27
26	16	29
27	17	31
28	19	33
29	20	36
30	22	38
31	24	40
32	25	42

d Compare to average 10 year risk

Probability		
age	women, %	men, %
30–34	<1	3
35–39	<1	5
40–44	2	6
45–49	5	10
50–54	8	14
55–59	12	16
60–64	13	21
65–69	9	30
70–74	12	24

These charts were prepared with the help of William B. Kannel, M.D., Professor of Medicine and Public Health and Ralph D'Agostino, PhD, Head, Department of Mathematics, both at Boston University; Keaven Anderson, PhD, Statistician, NHLBI, Framingham Study; Daniel McGee, PhD, Associate Professor, University of Arizona.

Efficiency of Current Risk Profiles

The current risk profiles evidently are not able to account for differences in the level of CHD incidence among population samples. However, they are relevant to distinguish between high- and low-risk persons within such populations. The ingredients of the coronary risk profile, such as the blood pressure, serum cholesterol and cigarette smoking, have been shown to be major risk factors in quite different sorts of human populations.

It is not easy to formulate an answer to the question as to how well the known risk factors for CHD account for the observed incidence with a population [19]. It is reasonable to make a judgement that the larger the ratio obtained by dividing the proportion of cases correctly designated at high risk by the proportion correctly designated at low risk, the better the risk function for prediction. By this criterion, the Framingham risk profile is quite useful because it defines risk over an extremely wide range.

It could be argued that the proportion of cases of CHD correctly designated at high risk is the preferred criterion for efficiency of prediction. However, this does not recognize that nobody is either certain to develop CHD or to be immune to it. Hence, the purpose of a risk formulation must be to provide a correct estimate of the probability. We can only strive to identify groups of persons with specified risk factors as having a certain probability of developing CHD as precisely as possible. All estimates have to be considered as confidential as the set of risk factors chose to make up the risk profile.

After 4 decades of epidemiologic research at the Framingham Study and elsewhere, the risk factor concept has become firmly established, so that CHD risk can now be predicted with reasonable accuracy from ordinary clinical office procedures. Categorical risk assessment according to the number of risk factors, arbitrarily defined, do also have utility and will identify high-risk persons. However, they tend to overlook the substantial proportion of the population at equally high risk because of multiple marginal abnormalities. It is important not to miss such persons. The multivariate coronary risk profile which quantitatively synthesizes risk factors and makes it possible to efficiently identify a small segment of the population from which a substantial proportion of the CHD events will evolve.

Long-Term versus Short-Term Predictors

Most of the major risk factors which predict long-term also predict short-term risk of coronary events and vice versa. [1]. Since thrombosis appears to be the immediate precipitating event in myocardial infarction and sudden death, one would expect fibrinogen to predict best in the short term. In fact, the Framingham Study and others show clearly that it predicts long-term risk factors as well [20]. Elevated white blood cell count has also been found to be predictive in the long-term. This too could be an indicator of active fissuring, inflamed lesions which should predict best in the short term, particularly since both are subject to great intraindividual variability.

Examination of the stability of risk factors within subjects over time in the Framingham Study indicates a surprising consistency of the major risk factors over time. Most people tend to stay in the same quintile of the risk factor distribution over 20 years. Use of the mean of a number of determinations of a risk factor provides a more precise prediction than a single determination, but only marginally better.

As examination of the long-versus short-term predictive capacity of the major risk factor in the Framingham Study indicates that LVH, intraventricular block or NSA ECG predict better in the short term than in the long term. However, they predict in the long term as well.

Epidemiologic Approach – Impact on Health Policy

It is fair to claim the prospective epidemiologic long-term studies, such as the Framingham Study, have provided important information over the past 4 decades that has helped researchers, scientists and physicians increase their understanding of the intrinsic and environmental factors predisposing to the major cardiovascular diseases. These contributors have been labelled 'risk factors', a term coined by the Framingham Study.

The Framingham Study legitimized the epidemiologic approach to investigation of causes of CHD. It provided information of CHD incidence, clues to pathogenesis and modifiable predisposing factors. By establishing the risk factors, it provided the means for public health measures to alter the ecology to one more favorable to cardiovascular health, gave to the public information needed to help them protect themselves and helped physicians to identify high-risk candidates for preventive measures requiring pharmaceuticals and skilled procedures on counselling.

The epidemiologic findings developed by long-term studies stimulated trials and then public health initiatives to control cigarette smoking in the 1960s, hypertension in the 1970s and serum cholesterol in the 1980s. The epidemiologic knowledge generated has influenced an increasing number of physicians to change their priorities from preoccupation with completed catastrophes to an emphasis on modifiable predisposing factors.

There is a growing appreciation that seemingly innocuous living habits and asymptomatic medical 'trivia' can have ominous health implications. Acceptable values for blood pressure, cholesterol and glucose have been revised downward, smoking rendered unacceptable, exercise promoted and eating habits revised.

Multivariate risk assessment instruments have been devised which enable physicians to more efficiently select candidates for prevention of CHD without either needlessly alarming or falsely reassuring too many patients. Only by using such risk profiles is it possible to control the number of persons who have to be treated to benefit one. With such tools at their disposal and effective preventive measures which are now available, physicians should come to regard the onset of overt CHD as a medical failure rather than the first indication for treatment.

References

1 Kannel WB: Contributions of the Framingham Study to preventive cardiology. J Am Coll Cardiol 1990; 15:206–211.
2 Kannel WB, McGee D, Gordon T: A general cardiovascular risk profile: The Framingham Study. Am J Cardiol 1976;38:56–51.
3 Truett J, Cornfield T, Kannel WB: A multivariate analysis of the risk of coronary heart disease in Framingham. J Chronic Dis 1967;20:511–524.
4 Walker SH, Duncan DB: Estimation of the probability of an event as a function of several independent variables. Biometrica 1967;54:167–179.
5 American Heart Association: Coronary Risk Handbook: Estimating the Risk of Coronary Heart Disease in Daily Practice. Dallas, American Heart Association, 1973.
6 Gordon T, Sorlie P, Kannel WB. Coronary heart disease, atherothrombotic brain infarction, intermittent claudication – A multivariate analysis of some factors related to their incidence: Framingham Study. 16-Year follow-up; in Kannel WB, Gordon T (eds): The Framingham Study: An Epidemiological Investigation of Cardiovascular Disease. Section 27, No 426-1301/1345. Washington, United States Government Printing Office, 1971.
7 Brittrim E: Probability of coronary heart disease developing. West J Med 1982;136:86–89.
8 Stampfer MJ, Sacks FM, Salvini S, Willet WC, Hennekens CH: A prospective study of cholesterol, lipoproteins and risk of myocardial infarction. N Engl J Med 1991;325: 373–381.
9 Castelli WP, Garrison RJ, Wilson PW, Abbott RD, Kalousdian S, Kannel WB: Incidence of coronary heart disease and lipoprotein cholesterol levels. The Framingham Study. JAMA 1986;256:2835–2838.

10 Kannel WB, Dannenburg AL, Levy D: Population implications of electrocardiographic left ventricular hypertrophy. Am J Cardiol 1987;60: 85–93.
11 Reaven GM: Insulin resistance and compensatory hyperinsulinemia: Role in hypertension, dyslipidemia and coronary heart disease. Am Heart J 1991;121:1283–1288.
12 Kannel WB, McGee DL, Castelli WP: Latest perspectives on cigarette smoking and cardiovascular disease: The Framingham Study. J Cardiac Rehabil 1984;4:276–277.
13 Schildkraut JM, Myers RH, Cupples LA, Kiely DK, Kannel WB: Coronary risk associated with age and sex of parental heart disease in the Framingham Study. Am J Cardiol 1989;63:555–559.
14 Anderson KM, Wilson PWF, Odell PM, Kannel WB: An updated coronary risk profile. A statement for health professionals. Circulation 1991;83:356–362.
15 Brand RJ, Rosenman RH, Sholtz RI, Friedman M: Multivariate prediction of coronary heart disease in the Western Collaborative Group Study compared to the findings of the Framingham Study. Circulation 1976;53:348–355.
16 Leaverton PE, Sorlie PD, Kleinman JC, Dannenberg AL, Ingster-Moore L, Kannel WB, Cornoni-Huntley JC: Representatives of the Framingham risk model for coronary heart disease mortality. A comparison with a national cohort study. J Chronic Dis 1987;40:775–784.
17 McGee D, Gordon T: The results of the Framingham Study applied to four other US-based epidemiologic studies of cardiovascular disease; in Kannel WB, Gordon T (eds): The Framingham Study: An Epidemiological Investigation of Cardiovascular Disease. Section 31, US Department of Health, Education and Welfare, No. 76-1083. Bethesda, United States Government Printing Office, 1976.
18 Pooling Project Research Group: Relationship of blood pressure, serum cholesterol, smoking habit, relative weight and ECG abnormalities to incidence of major coronary events: Final report. J Chronic Dis 1978;31:201–306.
19 Gordon T, Kannel WB: Multiple risk functions for predicting coronary heart disease: The concept, accuracy and application. Am Heart J 1982;103:1031–1039.
20 Kannel WB, Wolf PA, Castelli WP, D'Agostino RB: Fibrinogen and risk of cardiovascular disease. The Framingham Study. JAMA 1987; 258:1183–1186.

Cardiology 1993;82:153–171

M. Kornitzer
M. Dramaix
I. Beriot
M. Lannoy
H. Gheyssens
F. Kittel

Laboratory of Epidemiology and Social Medicine, School of Public Health, Brussels Free University, Brussels, Belgium

Twenty-Five-Year Mortality Follow-Up in the Belgian Bank Study

Key Words
All-cause mortality
Coronary heart disease
Risk factors
Prospective study, long-term

Abstract
Two cohorts of male bank employees aged 40–59 at baseline and free of coronary heart disease were pooled and followed during 25 years for all-cause and coronary mortality. The vital status was known for all 1,227 subjects, but in both banks, 8% of all deaths occurred without any clue to a possible cause. All-cause and coronary mortality were 36.9 and 10.5 per 100 subjects, respectively. In bivariate analysis (adjusting for age), systolic blood pressure, diastolic blood pressure, serum cholesterol and cigarette smoking predicted all-cause mortality. The same variables predicted coronary mortality. Body mass index was neither a predictor of all-cause nor of coronary mortality. In multivariate analysis, age, systolic blood pressure and cigarette smoking predicted independently all-cause mortality whereas the same variables and serum cholesterol predicted coronary mortality. For coronary mortality, the odds ratio (OR) of systolic blood pressure ≥ 160 against < 140 mm Hg was 2.68 (95% confidence interval, CI: 1.41–5.08). The OR for subjects with baseline serum cholesterol ≥ 260 versus < 260 mg/dl was 1.68 (95% CI: 1.31–2.51). The OR for cigarette smokers versus never smokers was 2.54 (95% CI: 1.39–4.64). Age-adjusted relative risk for all-cause mortality for subjects in the highest versus the lowest quintile of a multiple logistic function was 1.72 whereas for coronary mortality the relative risk was 4.94. These results confirm the long-term predictive power of the three major modifiable coronary risk factors when measured once in middle-aged male subjects.

M. Kornitzer
Campus Erasme, CP 597
808 Route de Lennik
B–1070 Brussels (Belgium)

0008–6312/93/
0823–0153$2.75/0

Introduction

Numerous prospective epidemiological studies, the first of them being the Framingham Study, have shown that besides age, the three major risk factors related to coronary heart disease (CHD) incidence and/or mortality are hypercholesterolemia, hypertension and cigarette smoking.

The predictive power of these risk factors has been observed during follow-ups ranging from 5 to 23 years [1–21]. However, publications on follow-up of 25 years or more are scarce [22–28]. Most of the studies which started in the 50s or at the beginning of the 60s concerned middle-aged males or females. Those subjects who survived have now reached old age, ranging from 65 to 90 years and older.

A fundamental question that these long-term studies can answer concerns the predictive power of the coronary risk factors over a long time period and in elderly individuals. In other words: are middle-aged subjects free from CHD but with coronary risk factors 'labelled' as being at increased risk for myocardial infarction or sudden death at a risk for dying from CHD over a long follow-up period? Moreover, are these same risk factors also long-term predictors of all-cause mortality? Answers to these questions would have important implications both for public health and clinical practice.

This paper reports 25-year mortality follow-up data from the first Belgian long-term prospective study, which started in the mid-60s: The Belgian Bank Study. Five- and 10-year incidence results of this study have been previously reported [29, 30].

Subjects and Methods

The baseline screening in 1964–1965 concerned 538 males aged 40–59 working in a private bank (participation rate: 95%) and 877 civil servant males working in a savings bank (participation rate: 91%). All subjects were employees, either clerks or executives. The following variables were measured in study participants.

In one cohort, blood pressure was measured in the sitting position, whereas in the other cohort, measures were done while subjects were recumbent. In both cohorts, a mercury sphygmomanometer was used. In one cohort, height and weight were measured with shoes on, in the other without shoes. Obesity was expressed by the body mass index (BMI): weight (in kg)/height (in m^2). All serum cholesterol samples were analyzed in Naples (Laboratory of Prof. Fidansa) using the method of Abell et al. [31]. For smoking habits, the same standardized questionnaire was used in both cohorts. The Rose questionnaire was administered to both cohorts to assess angina. All ECGs were coded according to the Minnesota code by one of the authors (M.K.).

In summary, measurement of blood pressure, weight and height were not standardized in the cohorts. However, 25-year all-cause and coronary mortality were similar in both cohorts as well as their relation with baseline, blood pressure, weight and height. Hence, the two cohorts were pooled for the 25-year follow-up results reported in this paper.

Follow-Up

Through the medical departments of both banks, vital status of all participants was periodically checked: as all employees were affiliated to a special retirement fund, death of any member of the staff of these banks was immediately reported to the personnel and medical departments. They, in turn, reported any of these deaths to us. For all deaths, records from their general practitioners or hospitalizations, when death occurred in a hospital, were obtained. These records were reviewed by one of the authors (M.K.). Death was attributed to CHD when there was a documented myocardial infarction or sudden and unexpected death without any evidence of another diagnosis. Sudden and unexpected death was defined as death occurring in less than 2 h or in subjects found dead when death occurred during their sleep. The 25-year follow-up for vital status was complete. In both banks, 8% of all deaths occurred without any suggestions as to a possible cause.

Subjects having a diagnosis of CHD at the time of the baseline examination (either angina, history of myocardial infarction or Minnesota codes $I_{1\text{-}2\text{-}3}$ or $IV^{1\text{-}2\text{-}3}$ or $V_{1\text{-}2\text{-}3}$ or VII_1) are excluded in this report. Thus, the follow-up concerns data from 1,227 subjects free from CHD at initial screening.

Statistics

All analyses concerning coronary mortality were done comparing the group dying from CHD to that of the subjects still alive or those dying from other causes. t tests and χ^2 tests were used when comparing means or for the analysis of contingency tables, respectively.

The confidence intervals (CI) of the relative risk (RR) have been obtained by the statistical program Statcalc from Epi-Info, version 5 (Taylor series 95% CI) [32]. Age adjustment of the RR has been done by stratification. The adjusted RR is obtained by the estimation of Mantel-Haenszel, and the CI are those of Greenland and Robins [33]; the statistical significance is that of the χ^2 of Mantel-Haenszel.

Survival curves have been estimated by the actuarial method, and the null hypothesis of no difference between two or more survival curves has been tested by the Lee and Desu test [34].

Multivariate analyses have been performed using the logistic regression. We have chosen logistic regression instead of any other method, such as Cox regression, in order to keep the same methods as those we used before in this study to analyze 10- and 15-year incidence rates, so that comparisons with preceding results are easier. Within certain models, the risk factors have been introduced as 'dummy variables' with the category corresponding to the lowest risk of death utilized as a reference in order to estimate the adjusted odds ratio (OR) as well as accompanying CI.

Quintiles have been formed from the distribution of the individual values of the logistic function. In order to take into account important age differences between quintiles, age-standardized mortality ratios (SMR) have been calculated taking the mortality rate for age of the whole sample as the reference. Finally, SMR for the four highest quintiles compared to the first quintile were obtained.

With the exception of the analysis of any Mantel-Haenszel's test of statistical significance, all the statistical analyses were done on the SPSS program, version 4.0 – Unix [35, 36].

Results

There was a total of 453 deaths over the follow-up period, of which 129 could be attributed to CHD (28.5%). Summarizing deaths from CHD, cerebrovascular disease and other cardiovascular diseases, total cardiovascular mortality was responsible for 45% of all deaths. Twenty-five-year all-cause mortality was 36.9/100 subjects while coronary mortality was 10.5/100 subjects.

Table 1 shows mean baseline risk factors in subjects alive, in those who died from any cause and in those whose death was attributed to CHD. For subjects who died versus subjects alive after 25 years of follow-up, age is significantly higher, and age-adjusted average systolic blood pressure, diastolic blood pressure and serum cholesterol are significantly higher; age-adjusted prevalence of cigarette smoking is also significantly higher in those who died. No significant differences were observed in baseline obesity as expressed by the BMI. When comparing those dying from CHD with those still alive, significant differences were observed for age; age-adjusted average systolic and diastolic blood pressure were 7.7 and 4.6 mm Hg higher in coronary deaths, respectively. Baseline age-adjusted serum cholesterol was 13.2 mg higher, and age-adjusted prevalence of cigarette smoking was 6.4% higher in coronary deaths. No significant differences were observed for BMI.

Table 2 shows univariate analyses when variables were categorized. For all-cause mortality, the RR was 3.59 in those aged 55–59 compared to those aged 40–44. For both systolic and diastolic blood pressure, age-adjusted RR for those with the highest values are 1.54 and 1.72, respectively. Serum cholesterol is also related to all-cause mortality, those in the highest quartile having a 28% increased risk of dying compared to those in the lowest quartile; the age-adjusted RR of cigarette smokers versus nonsmokers is 1.51, whereas no relationship is observed between total mortality and tertiles of BMI. For coronary mortality, the same pattern is observed with RR around 2 for those having the highest values for systolic blood pressure, diastolic blood pressure and serum cholesterol ≥260 mg/dl. Moreover, there is a significant trend for coronary mortality as related to quartiles of serum cholesterol; the RR for cigarette smokers versus nonsmokers is 2.01. Again, no significant relationship is observed with tertiles of BMI.

Figure 1a–e shows 25-year survival in relation to age, serum cholesterol, systolic blood pressure, diastolic blood pressure and smoking habits. Figure 2a–e illustrates 25-year survival from coronary death. Large differences are observed in terms of age, systolic blood pressure and diastolic blood pressure, both for long-term survival and for survival from coronary death. As can be seen, current cigarette smokers had the lowest overall survival rates (total mortality) whereas survival from coronary death is worse in ex-smokers than in cigarette smokers.

Table 1. Mean baseline risk factors and mortality from all causes and CHD

	All-cause mortality					Coronary mortality				
	deaths	alive	difference	CI 95%	p value	deaths	alive	difference	CI 95%	p value
Age, years	51.9	47.4	4.5	3.8–5.2	<0.001	51.7	47.4	4.3	3.2–5.3	<0.001
SBP, mm Hg	134.5	125.8	8.7	6.4–11.0	<0.001	134.4	125.8	8.2	4.3–12.9	<0.001
Age adjusted	133.6	126.3	7.3		<0.001	133.6	125.9	7.7		<0.001
DBP, mm Hg	83.8	79.9	3.9	2.5–5.3	<0.001	83.4	79.9	3.5	1.0–6.0	<0.01
Age adjusted	83.8	79.9	3.9		<0.001	83.5	78.9	4.6		<0.01
Serum cholesterol, mg/dl	246.3	238.5	7.8	2.2–13.4	<0.01	253.4	238.5	14.9	6.5–23.3	<0.001
Age adjusted	245.0	239.3	5.7		<0.05	251.9	238.7			<0.01
Cigarette smoking, %	59.4	51.3	8.1	2.3–13.9	<0.01	63.3	51.3	12	2.7–21.3	<0.05
Age adjusted	62.1	49.4	12.7		<0.001	65.8	49.4	16.4		<0.01
BMI, kg/m²	24.7	24.6	0.1	0.2–0.5	NS	24.8	24.6	0.2	0.4–0.8	NS

SBP = Systolic blood pressure; DBP = diastolic blood pressure.

Table 2. Univariate analysis – 25 years mortality from all causes and CHD: variables in categories

	All-cause mortality					Coronary mortality				
	per 100 subjects	RR	age-adjusted RR	95% CI	p value	per 100 subjects	RR	age-adjusted RR	95% CI	p value
Total	36.9					10.5				
Age (years)										
40–44	17	1.00				5.5	1.00			
45–49	30.9	1.82		1.34–2.44	<0.001	8.8	1.60		0.88–291	NS
50–54	42.5	2.50		1.94–3.21	<0.001	11.5	2.09		1.25–3.47	<0.01
55–59	60.8	3.58		2.82–4.51	<0.001	17.0	3.09		1.92–4.97	<0.001
SBP (mm Hg)										
<140	31.4		1.00			8.8		1.00		
140–149	41.4		1.15	0.94–1.40	NS	10.5		1.06	0.64–1.73	NS
150–159	57.8		1.63	1.34–1.98	<0.001	13.3		1.34	0.74–2.40	NS
≥160	64.7		1.54	1.29–1.83	<0.001	24.7		2.11	1.37–3.26	<0.001
DBP (mm Hg)										
<80	29.2		1.00			9.3		1.00		
80–89	37.3		1.23	1.02–1.47	<0.05	11.0		1.15	0.77–1.71	NS
90–99	43.3		1.39	1.15–1.68	<0.01	8.0		0.80	0.48–1.34	NS
≥100	57.6		1.72	1.40–2.11	<0.001	20.2		1.92	1.19–3.09	<0.01
Serum cholesterol (mg/dl)										
<260	34.1		1.00			8.8		1.00		
≥260	43.7		1.19	1.03–1.38	<0.05	14.6		2.64	1.45–4.81	<0.001
Quartiles										
Q1 (<210)	35.0					7.2				
Q2 (210–240)	33.7					9.4				
Q3 (240–270)	35.9				<0.05	11.7				0.006[a]
Q4 (≥270)	44.0					14.1				
Smoking										
Nonsmokers	27.7		1.00			6.3		1.00		
Ex-smokers	37.2		1.21	0.91–1.62	NS	14.7		2.08	1.11–3.90	<0.01
Pipe/cigare	36.5		1.24	0.95–1.62	NS	6.7		1.01	0.49–2.08	NS
Cigarette	40.3		1.51	1.22–1.86	<0.001	12.2		2.01	1.21–3.36	<0.001
BMI (kg/m^2)										
Tertiles										
T1 (<23.4)	38.6		1.00			9.8		1.00		
T2 (23.4–26)	33.2		0.86	0.79–1.02	NS	10.3		1.05	0.54–1.65	NS
T3 (≥26)	39.4		1.02	0.88–1.23	NS	11.4		1.16	0.92–2.54	NS

SBP = Systolic blood pressure; DBP = diastolic blood pressure.

[a] χ^2 for trend.

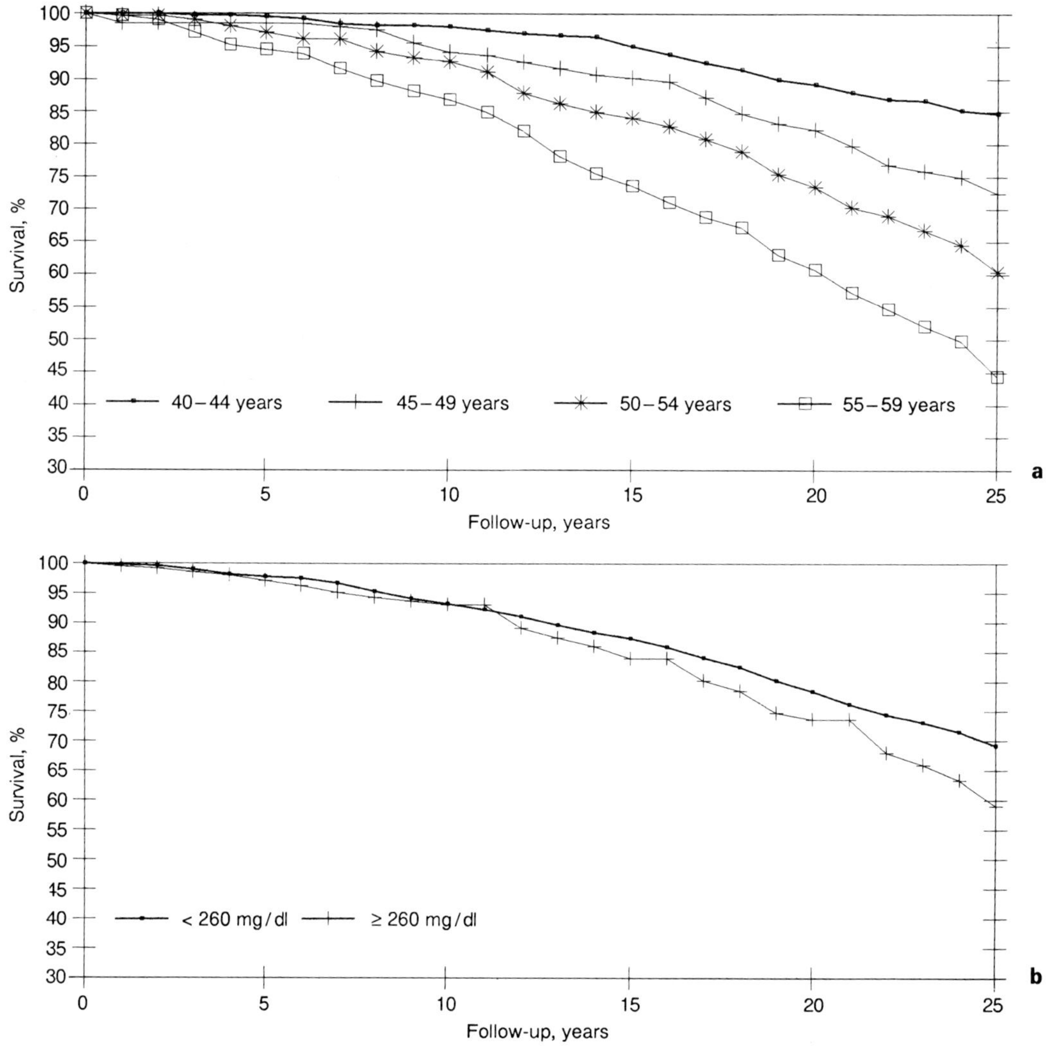

Fig. 1. 25-Year survival and age (**a**), serum cholesterol (**b**), systolic blood pressure (**c**), diastolic blood pressure (**d**) and smoking behavior (**e**) in subjects free of CHD at initial screening.

Table 3 shows the results of a multivariate analysis using a multilogistic model in which age, systolic blood pressure, serum cholesterol, BMI and smoking were simultaneously controlled for. The upper part of table 3 shows the logistic regression coefficients for all-cause mortality for three follow-up periods: 0–25 years, 0–10 years and 11–25 years. As can be seen, age and systolic blood pressure are significant independent predictors of all-cause mortality for all three follow-up periods; neither serum cholesterol nor BMI are independent predictors of all-cause mortality, although for the former, the 25-year follow-up

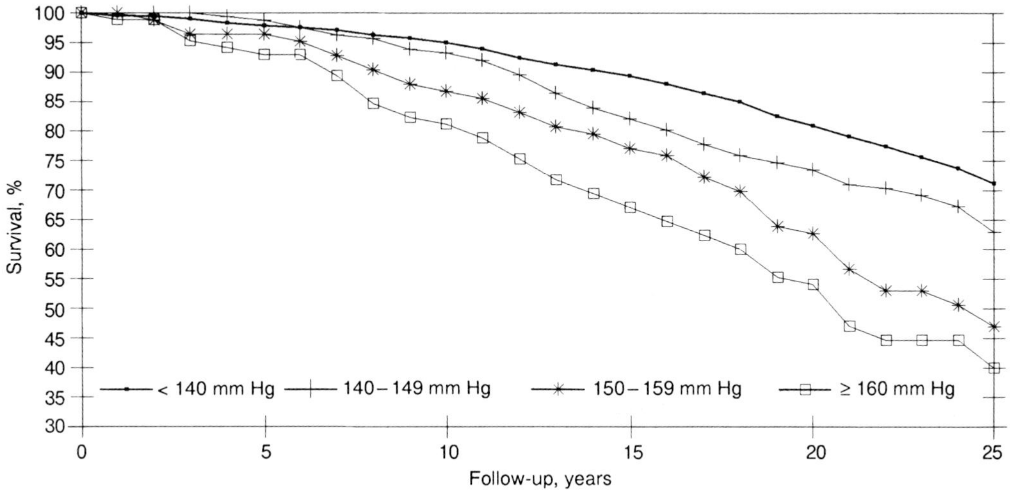

Survival, %
Follow-up, years
< 140 mm Hg
140–149 mm Hg
150–159 mm Hg
≥ 160 mm Hg

c

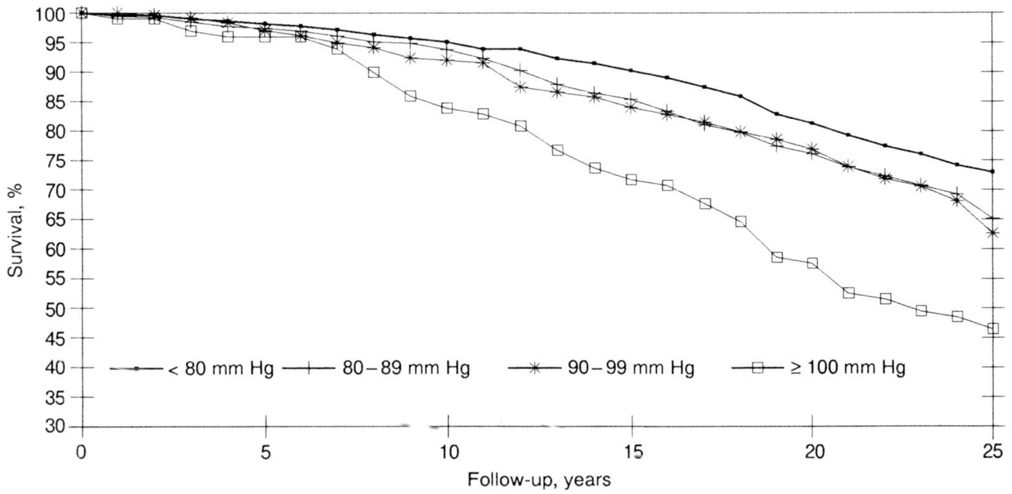

Survival, %
Follow-up, years
< 80 mm Hg
80–89 mm Hg
90–99 mm Hg
≥ 100 mm Hg

d

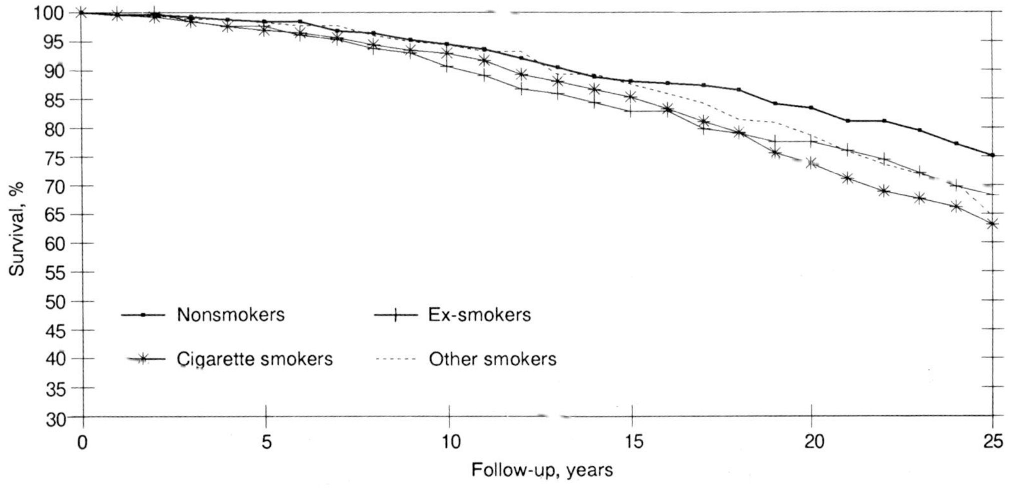

Survival, %
Follow-up, years
Nonsmokers
Ex-smokers
Cigarette smokers
Other smokers

e

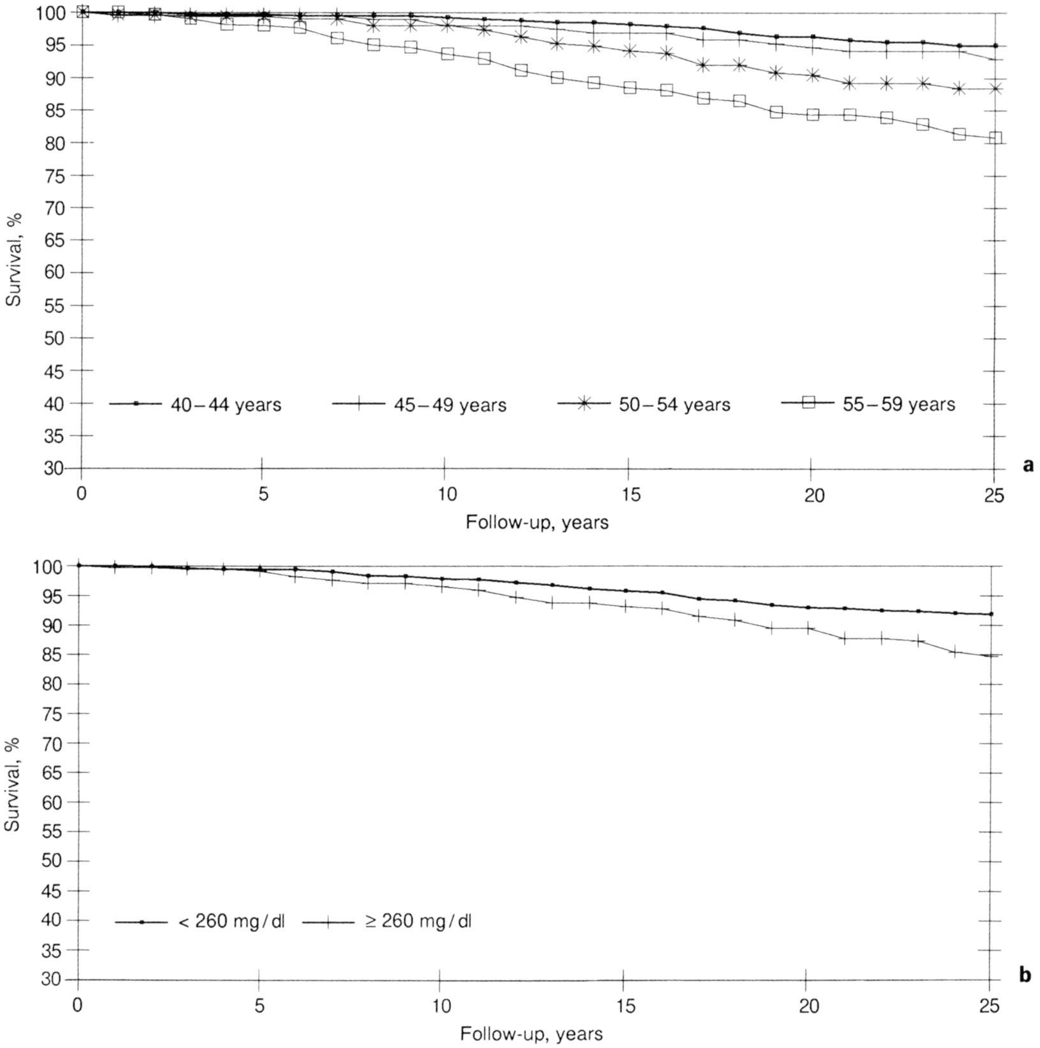

Fig. 2. 25-Year survival from coronary mortality and age (**a**), serum cholesterol (**b**), systolic blood pressure (**c**), diastolic blood pressure (**d**) and smoking behavior (**e**). Subjects dying from other causes were censored the year of their death.

period is of borderline statistical significance ($p = 0.097$). Cigarette smoking is an independent predictor for the complete 25-year follow-up period as well as for the last 15 years, but not for the first 10 years. The lower part of table 3 gives the logistic regression coefficients for coronary mortality: age, serum cholesterol and cigarette smoking are independent predictors for the 25-year follow-up, whereas systolic blood pressure is of a borderline statistical significance ($p = 0.057$). For the first 10 years of follow-up, age, systolic blood pressure and serum cholesterol are independent predictors, while for the last 15 years, age

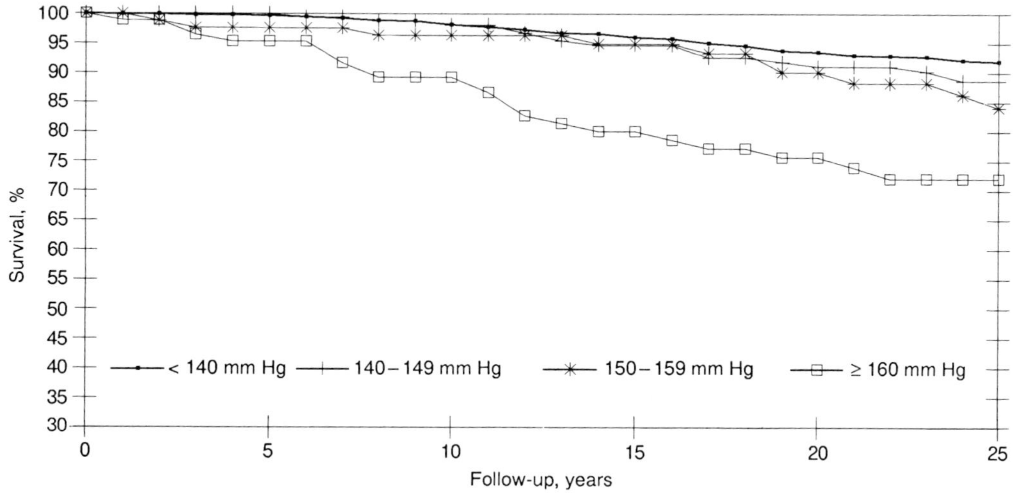
Survival, %
< 140 mm Hg
140–149 mm Hg
150–159 mm Hg
≥ 160 mm Hg
Follow-up, years

c

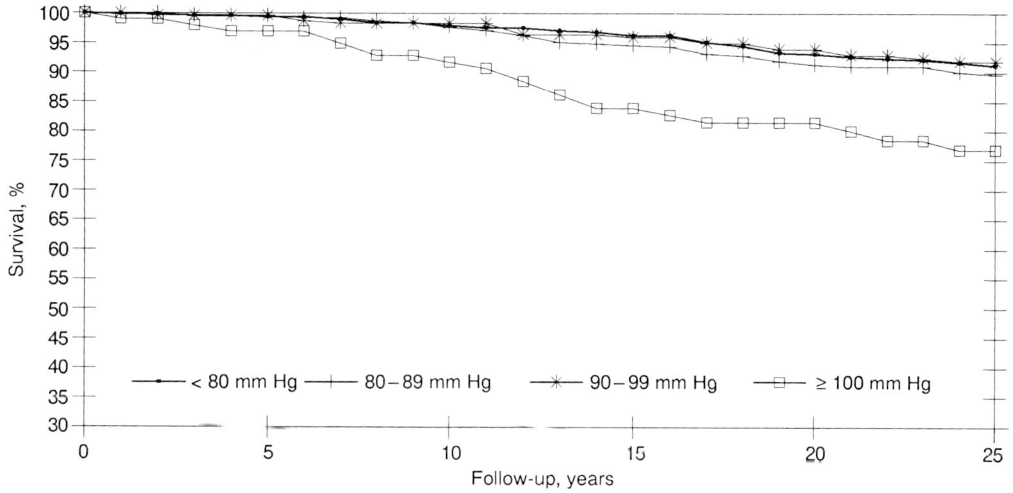
Survival, %
< 80 mm Hg
80–89 mm Hg
90–99 mm Hg
≥ 100 mm Hg
Follow-up, years

d

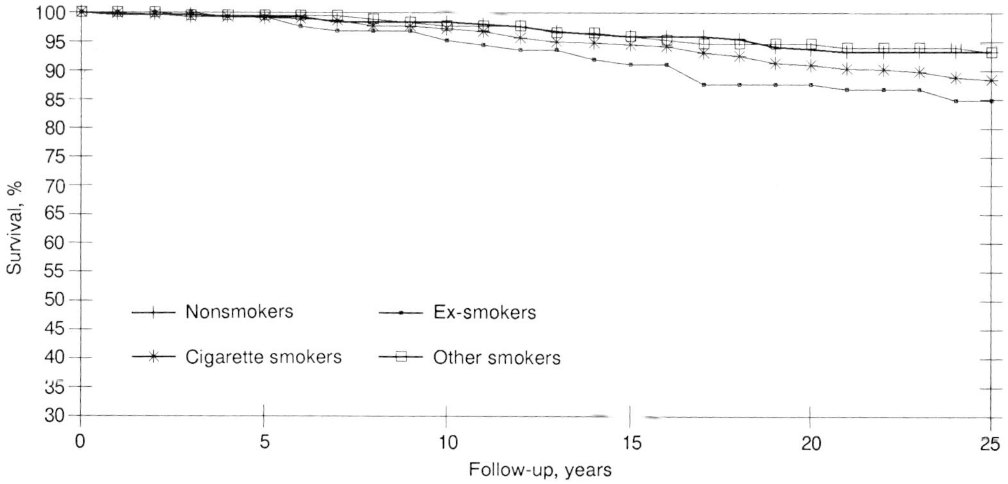
Survival, %
Nonsmokers
Ex-smokers
Cigarette smokers
Other smokers
Follow-up, years

e

Table 3. Multivariate analysis: logistic regression coefficients

Follow-up	Age	Systolic blood pressure	Serum cholesterol	BMI	Smoking
All-cause mortality					
0–25 years					
β coefficient	0.1274	0.229	0.0025	−0.0143	0.2578
SE	0.0121	0.0039	0.0015	0.0219	0.0590
p	<0.001	<0.001	0.097	0.514	<0.001
0–10 years					
β coefficient	0.1031	0.0192	−0.0002	−0.0198	0.1080
SE	0.0217	0.0054	0.0026	0.0354	0.0996
p	<0.001	<0.001	0.930	0.576	0.279
11–25 years					
β coefficient	0.1219	0.0215	0.0027	−0.0149	0.2687
SE	0.0129	0.0042	0.0017	0.0233	0.0634
p	<0.001	<0.001	0.115	0.521	<0.001
Coronary mortality					
0–25 years					
β coefficient	0.0702	0.0096	0.0057	−0.0070	0.2300
SE	0.0176	0.0050	0.0021	0.0311	0.0908
p	<0.001	0.057	0.007	0.823	0.011
0–10 years					
β coefficient	0.0993	0.0160	0.0085	−0.0208	0.1337
SE	0.0360	0.0085	0.0037	0.0573	0.1649
p	0.006	0.002	0.021	0.717	0.418
11–25 years					
β coefficient	0.0655	0.0088	0.0041	−0.0081	0.2516
SE	0.0200	0.0061	0.0026	0.0357	0.1060
p	0.001	0.149	0.113	0.821	0.014

and cigarette smoking are the only predictors. BMI is neither a predictor of all-cause nor coronary mortality.

When comparing expected and observed cases by quintile of the multiple logistic function (MLF) the correlation coefficient is 0.99 for all-cause mortality and 0.90 for coronary mortality.

Figures 3 and 4 depict 25-year cumulative all-cause and coronary mortality by quintiles of the MLF with age, systolic blood pressure, serum cholesterol, BMI and smoking controlled for. For all-cause mortality, there is a steady increase from quintile 1 to quintile 5, the RR of quintile 5 being 4.9 compared to quintile 1. For coronary mortality, there is an increase from the 1st to the 3rd quintile, then a slight decrease in the 4th quintile and finally, an important increase for quintile 5; the RR for quintile 5 as compared to quintile 1 is 10.2. In fact, there were only 4 coronary deaths in the 25-year follow-up for those in

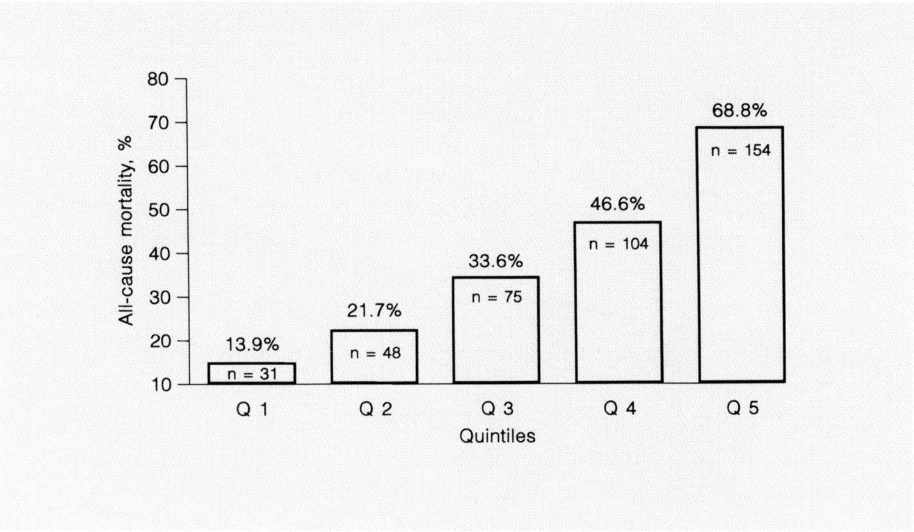

3

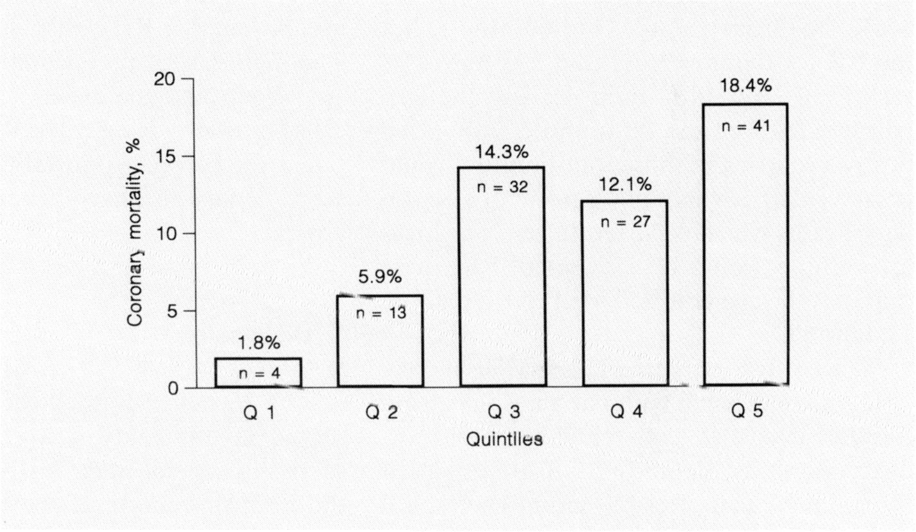

4

Fig. 3, 4. 25-Year cumulative all-cause and coronary mortality by quintiles of the MLF with age, systolic blood pressure, serum cholesterol, BMI and smoking controlled for. Forty-one (**3**) and 12 (**4**) deaths were not included because of missing values.

the 1st quintile of the MLF versus 41 coronary deaths for those in the 5th quintile. Part of those differences are due to mean differences in the modifiable risk factors across the quintiles of the MLF and part of the differences in age distribution: as an example, for coronary mortality, the average MLF are for the age groups 40–44 and 55–59 6.0 and 16.4%, respectively. Hence SMR for all-cause and coronary mortality give the differences

Table 4. SMR in quintiles of the MLF[a]

	Q1	Q2	Q3	Q4	Q5
All-cause mortality	100	121	126	137	172
Coronary mortality	100	100	277	369	494

[a] Quintile 1 as reference.

between quintiles only due to the modifiable coronary risk factors.

Table 4 lists SMR taking quintile 1 as reference. For all-cause mortality, SMR rises steadily from quintile 2 to quintile 5, but the RR for quintile 5/quintile 1 is now 1.72 instead of 4.95. For coronary mortality, the SMR rises from quintile 3 to quintile 5, but the RR for quintile 5/quintile 1 is now 4.94 instead of 10.2. If all quintiles had the age-adjusted MLF of the 1st quintile, 28% of all deaths occurring in those four quintiles could be prevented. Seventy-five percent of all coronary deaths which occurred in the four highest quintiles could be prevented, had these quintiles the age-adjusted MLF profile of the 1st quintile.

Table 5 gives a multivariate analysis by way of logistic regression with variables in categories. As can be seen, both for all-cause and coronary mortality, significant OR are observed for systolic blood pressure, serum cholesterol and cigarette smoking, independently from age, with these three modifiable risk factors being significantly predictive of both 25-year all-cause and coronary mortality.

When looking for low- or high-risk subjects at baseline, we found 33 subjects with a systolic blood pressure <120 mm Hg and serum cholesterol <220 mg/dl and never smokers; this small group was defined as the low-risk group (table 6). On the other hand, we found 19 subjects being smokers with serum cholesterol ≥ 260 mg/dl and systolic blood pressure ≥ 160 mm Hg and/or diastolic blood pressure ≥ 100 mm Hg: those subjects were hereby considered as a 'high-risk' group.

In table 6, the remainder of the population is compared to the low-risk group. The age-adjusted RR for all-cause mortality is 2.61; there were no coronary deaths in the low-risk group, hence a RR could not be computed. The high-risk group is compared to the low-risk group and the age-adjusted RR for all-cause mortality is 2.82. No RR could be computed for coronary mortality due to the absence of any coronary death in the low-risk group.

Discussion

The 36.9% 25-year all-cause mortality observed in our study is lower than that observed in several other long-term studies. In the Finnish cohorts of the Seven Countries Study, 25-year all-cause mortality was 55.7% [23] and in the Italian cohorts of that study, the figure was 47.5% [24]. This could be due to geographical differences, life-style characteristics and to the fact that our cohort consisted of an occupational group which probably had a favorable mortality experience due in part to the 'healthy worker effect'. The proportion of coronary deaths was almost identical in the Zutphen study and in ours, being 26 and 28.3%, respectively. In the Finnish cohort of the Seven Countries Study, 40% of all

Table 5. 25-Year mortality follow-up – multivariate analysis by way of logistic regression with variables in categories[a]

	All-cause mortality		Coronary mortality	
Systolic blood pressure (mm Hg)				
<140				
140–149	OR 2/1	1.38 (0.93–2.03)	OR 2/1	1.31 (0.74–2.33)
150–159	OR 3/1	3.00 (1.79–5.03)	OR 3/1	1.58 (0.89–2.81)
≥160	OR 4/1	3.19 (1.81–1.72)	OR 4/1	2.68 (1.41–1.62)
p value		<0.001		<0.05
Serum cholesterol				
(>260/≤260 mm Hg)	OR	1.37 (1.03–1.82)	OR	1.68 (1.13–2.51)
p value		<0.05		<0.05
Cigarette smoking				
(cigarette smokers/nonsmokers)	OR	2.10 (1.47–3.01)	OR	2.54 (1.39–4.64)
p value		<0.001		<0.01

[a] Variables in the equation: age, systolic blood pressure, serum cholesterol, cigarette smoking.

Table 6. Mortality in low-risk subjects

	Low-risk group (n = 33; 3%)		Rest of the population (n = 1,085; 93%)		Rest of the population/ low-risk group age-adjusted RR	High-risk group (n = 19; 2%)		High-risk/ low-risk group age-adjusted RR
	n	%	n	%		n	%	
All-cause mortality	7	21.2	406	37.4	2.61 (0.85–3.07)	13	68.4	2.82 (1.39–5.72)
Coronary mortality	0	0	118	10.9	–	5	26.3	–
Noncoronary mortality	7	21.2	288	26.5	1.14 (0.60–2.16)	8	42.1	1.73 (0.77–3.91)

Low-risk group: systolic blood pressure <120 mm Hg *and* serum cholesterol ≤220 mg/dl *and* never smokers; high-risk group: smokers *and* serum cholesterol ≥260 mg/dl *and* systolic blood pressure ≥160 mm Hg or/and diastolic blood pressure ≥100 mm Hg.

deaths were due to CHD [23], whereas in the Italian cohort of that study, the proportion was only 25.6% [24]. However, we have to take into account that in our study, a fair proportion of the 8% of deaths with unknown diagnoses were probably due to CHD.

An important question that long-term follow-up studies could answer concerns the predictability of the major coronary risk factors in terms of all-cause and coronary mortality or incidence. Age, serum cholesterol, systolic and diastolic blood pressure, BMI and cigarette smoking were the coronary risk factors looked for in relation to these outcomes, first in univariate and then in multivariate analyses.

Serum Cholesterol

In our study, age-adjusted baseline serum cholesterol is related to 25-year all-cause mortality in univariate analyses; using the multilogistic model, serum cholesterol is no longer a significant predictor of all-cause mortality after adjusting for other potentially confounding factors. In most of the reviewed studies, the level of serum cholesterol was not a predictor of all-cause mortality: this was the case in the 10-year follow-up of the Seven Countries Study with the exception of the American railroad cohort [8], the 20-year follow-up of the Italian cohorts of that study [18], the 23-year follow-up of the Israel Civil Servant Study [20], the 25-year follow-up of the Zutphen Study [22] and, finally, in the 12-year follow-up in the population survey in Renfrew and Paisley [5]. However, in the Framingham Study, overall mortality was significantly related to serum cholesterol only in those younger than 45 years at baseline [26] and, in the Whitehall Study, in those who died before the age of 60 [14].

On the other hand, there is a consistency between studies relating serum cholesterol to coronary mortality or incidence in both short and long periods of follow-up: this was the case in the Seven Countries Study [8], in the Whitehall Study [10], in the 7-year follow-up of the primary screenees of the MRFIT [2], in the Stockholm Prospective Study [15] and in the Italian rural cohorts of the Seven Countries Study followed during 25 years [25]. In the Whitehall Study, serum cholesterol predicted coronary mortality in elderly people at least until the age of 80, but the strength of the association fell with age at screening [14]. Serum cholesterol was predictive in studies with longer follow-ups: the Israel Civil Servants Heart Study [20], the Zutphen Study [22], the Finnish cohort of the Seven Countries Study [23] and in the Framingham Study [37].

Our results support all of these studies: serum cholesterol is a powerful independent predictor of 25-year CHD mortality, but the strength of this association is primarily observed during the first 10 years of follow-up. Our age-adjusted RR for those with cholesterol ≥ 260 mg/dl against those with cholesterol <260 mg/dl is 2.64 and close to the RR around 2 observed in most of the other reviewed studies. In summary, serum cholesterol is a long-term predictor of CHD mortality.

High Blood Pressure

We observed a strong predictive power of systolic and diastolic blood pressure for both all-cause and coronary mortality. In multivariate analyses, the association with all-cause mortality is stronger than for CHD mortality.

The relation of high blood pressure to all-cause mortality was also observed in all the cohorts of the Seven Countries Study [8], in a 20-year follow-up of the two rural Italian villages of that study [25], in the Framingham Study [27] and in the Men Born in 1913 Study in Göteborg [11].

In summary, systolic as well as diastolic blood pressure are long-term predictors of all-cause mortality; on average, those in the highest quintile of the blood pressure distribution have at least twice the risk of those in the lowest quintile.

With regard to CHD, there is consistency in the results of numerous prospective studies. Over the short-term follow-up, both systolic blood pressure and diastolic blood pressure predict CHD incidence and mortality in middle-aged males [4, 8, 15, 27, 28]. In the Whitehall Study, the prediction of 10-year coronary mortality was better for systolic than for diastolic blood pressure: standardized RR of systolic blood pressure, adjusted for diastolic blood pressure, was 1.29, and for diastolic blood pressure adjusted for systolic blood pressure, it was only 1.09 [9]. In the Finnish cohorts of the Seven Countries Study, the hazard ratio for systolic blood pressure in the highest against the lowest quintile was 2.44 [23].

Our study confirms the long-term predictive power of systolic blood pressure both for all-cause and coronary mortality.

Smoking Behavior

In our study, cigarette smoking is an independent predictor of 25-year all-cause and coronary mortality. The predictive power of cigarette smoking for all-cause mortality was observed in several population studies [8, 17, 18, 38] as well as in male British doctors [19]. The relation of cigarette smoking to CHD was observed in several studies in both the United States [17, 38] and in Europe [23, 24].

In summary, cigarette smokers are at higher risk for both all-cause and coronary mortality. Comparing cigarette smokers to never smokers, the long-term RR is around 2.

We observed a significantly higher risk for coronary mortality in baseline ex-smokers compared to never smokers with an age-adjusted RR of 2.08. Most of the studies report a lower risk in ex-smokers compared to smokers: this was observed in the 10-year follow-up of the northern European cohorts, the Yugoslav cohorts and the American railroad cohorts of the Seven Countries Study [8]. The CHD risk decreased within a short period in the Framingham Study [39]. However, some studies failed to observe significant differences between ex- and current smokers. This was the case in the Western Collaborative Group Study after 4.5 years of follow-up [40], in the southern European cohorts of the Seven Countries Study [8] and in the 20-year follow-up of the British Doctors Study [19]. These discrepancies are probably explained by information bias at baseline, some ex-smokers being in fact current smokers due to misclassification during follow-up. These hypotheses could not be verified in the absence of routinely used biological markers like cotinine, thiocyanate or carbon monoxide at the time of the baseline screening in these studies.

Obesity

Using BMI as an indicator of obesity, we did not observe any significant association with either total or coronary mortality, in univariate or in multivariate analyses.

Having reviewed the subject [41], we can summarize that in studies ranging from 5 to 14 years of follow-up, no independent relationship was found between indices of obesity and incidence of myocardial infarction or CHD mortality [7, 8, 12, 42, 43] with some exceptions. In the Framingham Study, sudden cardiac death was independently predicted by relative weight after 12 years of follow-up [44]. In the Zutphen study, the age-adjusted BMI predicted 10-year coronary mortality [8]. Longer follow-up studies reported mixed results. In the Manitoba Study [45], BMI predicted 26-year coronary mortality (in fact, only sudden death), while in the Framingham Study [46], relative weight independently predicted 26-year incidence of CHD in both males and females. On the other hand, BMI did not predict either 20-year all-cause mortality in the two rural Italian cohorts from the Seven Countries Study [18] or 25-year coronary mortality in the Italian, Dutch and Finnish cohorts of the Seven Countries Study [24].

Forty years ago, Vague [47] suggested that differences in regional adiposity were correlated with different metabolic consequences on both lipid and carbohydrate metabolism. In fact, only the upper body or 'masculine' adiposity was related to several atherogenic factors. This hypothesis has been recently confirmed in several prospective studies [48–52]. There seems to be no straightforward explanation for the contradictory results observed between North American and European cohorts concerning the long-term predictive power of total obesity, as measured by relative weight or the BMI. The spreads of the distributions may well be different with more extreme obese subjects in the North American studies or, alternatively, differences in the prevalence of upper body or masculine obesity in North American as compared to European populations.

Multivariate Approach

In combining the examined risk factors in a comprehensive coronary risk factor profile according to the MLF the discriminatory power of the main coronary risk factor was shown for both all-cause and coronary mortality. In our study, when standardizing for age the modifiable coronary risk profile is about 70% higher in the 5th compared to the 1st quintile for all-cause mortality and about 5 times higher in the 5th quintile compared to the 1st quintile for coronary mortality.

Pekkanen et al. [23] observed that age-adjusted 25-year coronary mortality in the most favorable group was only 12 against 75% in those with the worst coronary risk profile or a RR of 6, which is close to our age-adjusted RR of 4.95.

When follow-up is divided into two periods, 0–10 and 11–25 years, we observed the following: (1) *for all-cause mortality,* systolic blood pressure was an independent predictor during both periods, serum cholesterol in neither of them, and cigarette smoking was only an independent predictor from the 11th year on, and (2) *for coronary mortality,* systolic blood pressure and serum cholesterol were both independent predictors during the first period while smoking appeared as a predictor from the 11th year on.

In the Zutphen Study [22], serum cholesterol was only an independent predictor of CHD mortality during the first 15 years of follow-up, but not in the following decade. In the Whitehall Study, the RR for coronary mortality (for a difference of 76 mg/dl in serum cholesterol) seemed to increase with the interval of follow-up, but the trend was marginally significant. However, in the 25-year follow-up of the Finnish cohorts, the relationship of serum cholesterol with risk of CHD mortality remained almost unchanged over time [23]. There is no ready explanation for the difference observed in that study with both the Zutphen Study and ours. The average serum cholesterol is much higher in Finland than in The Netherlands and Belgium. A better tracking of high-risk subjects may well be related to serum cholesterol distribution in the population. We can also speculate that the prevalence of subjects at high risk during initial survey is related to the prevalence of subclinical atherosclerosis. Finally, the decrease in coronary mortality observed during the seventies was steeper in the Dutch and Belgian populations than in the Finnish one. The same hypothesis could be invoked for the predictive power of high blood pressure: the 25-year prediction of coronary mortality was mainly due to its predictive power during the first 10 years of follow-up.

Our observation that cigarette smoking predicted CHD mortality from the 11th year on is not easy to explain: selective survival, changes in smoking habits or other factors could explain the findings observed. As already pointed out, most of the short-term follow-up studies reported a significant independent relation of CHD incidence or mortality with cigarette smoking. In the Finnish Study

[23], the predictive power decreased with time. If indeed misclassification occurred in our study with former heavy cigarette smokers relapsing during follow-up, the proportion of heavy smokers in the cohort could eventually increase over time, increasing the risk of cigarette smokers compared to never smokers. Another hypothesis could be that the average daily number of cigarettes smoked was lower in our study than in others and that, consequently, expressed in cigarette packs/year, the years under risk needed to be longer in order to observe a significant difference between smokers and nonsmokers in terms of all-cause and coronary mortality.

Given the probability of misclassification or regression dilution bias [53] due to the fact that subjects are classified after only one measure of serum cholesterol or blood pressure at the initial survey and, on the other hand, of a probable selective removal by death during the first years of follow-up for the most susceptible subjects at the upper tail of the distribution of serum cholesterol or blood pressure, the long-term prediction of the three major modifiable coronary risk factors strengthens the view that the detection and treatment of these risk factors should start in middle-aged subjects. It is hoped that interventions on these risk factors would result in a reduced risk of both all-cause and coronary mortality.

References

1 Keys A: Coronary heart disease in seven countries. Am Heart Assoc Monograph 29, 1970.

2 Stamler J, Wentworth D, Neaton J: Is relationship between serum cholesterol and risk of premature death from coronary heart disease continuous and graded? Findings in 356 222 primary screenees of the Multiple Risk Factor Intervention Trial (MRFIT). JAMA 1966;28:2823–2829.

3 Kannel W, Neaton J, Wentworth D, Thomas H, Stamler J, Hulley S, Kjelsberg M, for the MRFIT Research Group: Overall and coronary heart disease mortality rates in relation to major risk factors in 325 348 men screened for the MRFIT. Am Heart J 1986;112:825–836.

4 Martin M, Hulley S, Browner W, Kuller L, Wentworth D: Serum cholesterol, blood pressure and mortality: Implications from a cohort of 361 662 men. Lancet 1986;ii:933–936.

5 Isles C, Hole D, Gillis C, Hawthorne V, Lever A: Plasma cholesterol, coronary heart disease and cancer in the Renfrew and Paisley Survey. BMJ 1989;298:920–924.

6 Kozarevicz Dj, McGee D, Vojvodic N, Gordon T, Racic Z, Zukel W, Dawber T: Serum cholesterol and mortality: The Yugoslavia Cardiovascular Disease Study. Am J Epidemiol 1981;114:21–28.

7 Rosenman R, Brand R, Sholtz R, Friedman M: Multivariate prediction of coronary heart disease during 8.5-year follow-up in the Western Collaborative Group Study. Am J Cardiol 1976,337:903–910.

8 Keys A: Seven countries: A Multivariate Analysis of Death and Coronary Heart Disease. Cambridge, Harvard University Press, 1980.

9 Lichtenstein MJ, Shipley MJ, Rose G: Systolic and diastolic blood pressure as predictors of coronary heart disease mortality in the Whitehall Study. Br Med J 1985;291:243–245.

10 Rose G, Shipley M: Plasma cholesterol concentration and death from coronary heart disease: 10 years results of the Whitehall Study. Br Med J 1986;293:306–307.

11 Samuelsson O, Wilhelmsen L, Svärdsud K, Pennert K, Wedel H, Berglund G: Mortality and morbidity in relation to systolic blood pressure in two populations with different management of hypertension: The study of men born in 1913 and the Multifactorial Primary Prevention Trial. J Hypertens 1987;5:57–66.

12 Tibblin G, Wilhelmsen L, Werkö I. Risk factors for myocardial infarction and death due to ischemic heart disease and other causes. Am J Cardiol 1975;35:514–522.

13 Mariotti S, Capocaccia R, Farchi G, Menotti A, Verdecchia A, Keys A: Age, period, cohort and geographical area effects on the relationship between risk factors and coronary heart disease mortality: 15-year follow-up of the European cohorts of the Seven Countries Study. J Chronic Dis 1986;39:229–242.

14 Shipley M, Pocock S, Marmot M: Does plasma cholesterol concentration predict mortality from coronary heart disease in elderly people? 18 year follow-up in the Whitehall Study. BMJ 1991;303:89–92.

15 Carlson L, Böttiger L: Risk factors for ischaemic heart disease in men and women: Results of the 19-year follow-up of the Stockholm Prospective Study. Acta Med Scand 1985; 218:207–211.
16 Goldburt U, Holtzman E, Neufeld H: Total and high density lipoprotein cholesterol in the serum and risk of mortality: Evidence of a threshold effect. Br Med J 1985;290: 1239–1243.
17 Kannel W: Update on the role of cigarette smoking in coronary artery disease. Am Heart J 1981;101:319–328.
18 Menotti A, Conti S, Dima F, Giampoli S, Giuli B, Rumi A, Seccareccia F, Signoretti P: Prediction of all causes of death as a function of some factors commonly measured in cardiovascular population surveys. Prev Med 1983;12:318–325.
19 Doll R, Peto R: Mortality in relation to smoking: 20 years' observations on male British doctors. Br Med J 1976;ii:1525–1536.
20 Goldburt U, Yaari S: Cholesterol and coronary heart disease mortality: A 23-year follow-up study of 9902 men in Israel. Arteriosclerosis 1990;10:512–519.
21 Keys A, Taylor H, Blackburn H, Brozek J, Anderson J, Simonson E: Mortality and coronary heart disease among men studied for 23 years. Arch Intern Med 1971;128: 201–214.
22 Kromhout D, Bosschieter EB, Drijver M, Coulander C de L: Serum cholesterol and 25-year incidence of and mortality from myocardial infarction and cancer: The Zutphen Study. Arch Intern Med 1988;148:1051–1055.
23 Pekkanen J, Nissinen A, Puska P, Punsar S, Karvonen MJ: Risk factors and 25 year risk of coronary heart disease in a male population with a high incidence of the disease: The Finnish cohorts of the Seven Countries Study. BMJ 1989;299: 81–85.
24 Menotti A, Keys A, Kromhout D, Nissinen A, Blackburn H, Fidanza F, Giampaoli S, Karvonen MJ, Pekkanen J, Punsar S, Seccareccia F: Twenty-five-year mortality from coronary heart disease and its prediction in five cohorts of middle-aged men in Finland, The Netherlands and Italy. Prev Med 1990;19: 270–278.
25 The Italian Research Group of the Seven Countries Study: Twenty-five year incidence and prediction of coronary heart disease in two Italian rural population samples. Acta Cardiol 1986;41:283–299.
26 Anderson KM, Castelli WP, Levy D: Cholesterol and mortality: 30 years of follow-up from the Framingham Study. JAMA 1987;257: 2176–2180.
27 Stokes J, Kannel W, Wolf P, Cupples L, D'Agostino R: The relative importance of selected risk factors for various manifestations of cardiovascular disease among men and women from 35 to 64 years old: 30 years of follow-up in the Framingham Study. Circulation 1987; 75(suppl 5):65–73.
28 Stokes J III, Kannel W, Wolf P, D'Agostino R, Cupples L: Blood pressure as a risk factor for cardiovascular disease: The Framingham Study – 30 years of follow-up. Hypertension 1989:13(suppl I):13–18.
29 Kornitzer M, Thilly C, Vanroux A, Balthazar E: Incidence of ischaemic heart disease in two cohorts of Belgian clerks. Br J Prev Soc Med 1975; 29:91–97.
30 Kornitzer M, Dramaix M, Gheyssens H: Incidence of ischemic heart disease in two Belgian cohorts followed during 10 years. Eur J Cardiol 1979;9:455–472.
31 Abell L, Levy B, Bradie B, Kendall F: A simplified method for the estimation of total cholesterol in serum and the demonstration of its specificity. J Biol Chem 1952;195:357–359.
32 Epidemiology Program Office: Epi-Info. Atlanta, Centers for Disease Control, 1990.
33 Greenland S., Robins JM: Estimation of a common effect parameter from sparse follow-up data. Biometrics 1985;41:55–68.
34 Lee E, Desu M: A computer program for comparing κ samples with right censored data. Comput Programs Biomed 1972;2:315–321.
35 Marijä J: Statistical package for the social sciences. Norussis/SPSS Inc., 1990.
36 Armitage P, Berry G: Statistical Methods in Medical Research, ed 2. Oxford, Blackwell Scientific Publications, 1987.
37 Kannel W: Cholesterol and risk of coronary heart disease and mortality in men. Clin Chem 1988;34: B53–B59.
38 Friedman G, Dales L, Ury K: Mortality in middle-aged smokers and non-smokers. N Engl J Med 1979; 300:213–217.
39 Gordon T, Kannel W, McGee D: Death and coronary attacks in men after giving up cigarette smoking: A report from the Framingham Study. Lancet 1974;ii:1345–1348.
40 Jenkins CD, Rosenman RH, Zyzanski SJ: Cigarette smoking: Its relationship to coronary heart disease and related risk factors in the Western Collaborative Group Study. Circulation 1968;38:1140–1155.
41 Kornitzer M, Dramaix M: Obesité, adiposité et risque de cardiopathie ischémique: une mise au point. J Cardiol 1991;4:152–165.
42 Dyer A, Stamler J, Berkson D, Ländberg H: Relationship of relative weight and body mass index to 14-year mortality in the Chicago peoples gas company study. J Chronic Dis 1975;28:109–123.
43 Rose G, Reid D, Hamilton P, McCartney P, Keen H, Jarrett R: Myocardial ischaemia, risk factors and death from coronary heart disease. Lancet 1977;309:105–109.
44 Kannel W, LeBauer E, Dawber T, McNamara P: Relation of body weight to development of coronary heart disease. The Framingham Study. Circulation 1967;35:734–744.
45 Rabkin SW, Mathewson FAI, Hsu PH: Relation of body weight to development of ischemic heart disease in a cohort of young North-American men after a 26-year observation period: The Manitoba Study. Am J Cardiol 1977;39:452–458.

46 Hubert HB, Feinleib M, McNamara PM, Castelli WP: Obesity as an independent risk factor for cardiovascular disease: A 26-year follow-up of participants in the Framingham heart study. Circulation 1983;67: 968–977.

47 Vague J: Degree of masculine differentiation of obesities: Factors determining predisposition to diabetes, atherosclerosis, gout and uric calculous disease. Am J Clin Nutr 1956;4: 20.

48 Larsson B, Svärdsudd K, Welin I, Wilhelmsen I, Björntorp P, Tibblin G: Abdominal adipose tissue distribution, obesity and risk of cardiovascular disease and death: 13 year follow-up of participants in the study of men born in 1913. Br Med J 1984;288:1401–1404.

49 Lapidus I, Bengtsson C, Larsson B, Pennert K, Rybo E, Sjöström I: Distribution of adipose tissue and risk of cardiovascular disease and death: A 12-year follow-up of participants in the population study of women in Gothenburg, Sweden. Br Med J 1984;289:1257–1261.

50 Ducimétière P, Richard J, Cambien F, Avons P, Jacqueson A: Relationships between adiposity measurements and the incidence of coronary heart disease in a middle-aged male population: The Paris Prospective Study; in Vague J, et al. (eds): Metabolic Complications of Human Obesities. Amsterdam, Elsevier Science Publishers, 1985.

51 Kannel W, Cupples A, Ramaswami R, Stokes J III, Kreger B, Higgins M: Regional obesity and risk of cardiovascular disease: The Framingham Study. J Clin Epidemiol 1991;44: 183–190.

52 Higgins M, Kannel W, Garrison R, Pinsky J, Stokes J III: Hazards of obesity – The Framingham experience. Acta Med Scand Suppl 1988; 723:23–36.

53 MacMahon S, Peto R, Cutler J, Collins R, Sorlie P, Neaton J, Abbot R, Godwin J, Dyer A, Stamler J: Blood pressure, stroke, and coronary heart disease. 1. Prolonged differences in blood pressure: Prospective observational studies corrected for the regression dilution bias. Lancet 1990; 335:765–774.

Cardiology 1993;82:172–180

A. Menotti
F. Seccareccia
M. Lanti
S. Giampaoli
F. Dima

Laboratory of Epidemiology and Biostatistics, Istituto Superiore di Sanità, Rome, Italy

Time Changes in Predictability of Coronary Heart Disease in an Italian Aging Population

Key Words
Risk factors
Predictability, coronary events
Aging

Abstract
A pool of two Italian rural population samples made up of 1,712 men aged 40–59 at entry was studied in 1960 and than followed up for 25 years. The multivariate analysis of the first major coronary event using the Cox model showed, in men aged 40–59, 45–64 and 50–69, the significant predictive role of age, systolic blood pressure, serum cholesterol and cigarette smoking, but not of body mass index, without marked differences attributable to the aging process. Changes in systolic blood pressure, serum cholesterol and cigarette smoking occurring between year 0, 5 and 10 of follow-up, as defined by two different indicators, increased significantly the predictability of coronary events occurring between years 10 and 25 of follow-up when added to the model including the baseline factors. It is inferred that increases and decreases (even of relative nature) of the three major risk factors around the entry levels are associated with higher and lower levels of coronary risk in the 15 years after the changes have occurred.

Introduction

Epidemiological studies conducted longitudinally on population samples allow to confirm the role of well-established risk factors in the prediction of coronary heart disease (CHD) [1–7] in different population settings and geographical areas. When the follow-up becomes particularly long, some more specific questions can be asked, i.e. (1) whether single measurements of risk factors are predictive of events for different lengths of follow-up; (2) whether subsequent measurements of risk factors in aging cohorts are equally predictive, and (3) whether changes in risk factors in given periods of time contribute to the predic-

Research partly developed within the Italian National Research Council (CNR) Targeted Project 'Prevention and Control Disease Factors', Subproject 'Community Medicine', Contract 91.00289.PF41.

Prof. Alessandro Menotti
Laboratory of Epidemiology and Biostatistics
Istituto Superiore di Sanità
Viale Regina Elena 299
I-00161 Rome (Italy)

0008–6312/93/
0823–0172$2.75/0

tion of the events in subsequent time periods [8–18].

In Italy, within the Seven Countries Study on Cardiovascular Diseases [2, 19–22], three cohorts of men aged 40–59 were enrolled in 1960–1962, but only two of them, i.e. the rural ones, were monitored for both fatal and nonfatal coronary events up to 25 years of follow-up (only mortality, after the 10th year, for the third one).

Data on CHD incidence and mortality and on their prediction have been published elsewhere [23–29], but the three questions mentioned above were never tackled in a systematic way.

Material and Methods

The study population includes a pool of two rural samples enrolled and first examined in 1960. They comprised men aged 40–59 residing in defined geographical-administrative areas corresponding to the rural municipalities of Crevalcore in Northern Italy and of Montegiorgio in Central Italy. Within the enrolled samples, 993 men were examined in Crevalcore and 719 in Montegiorgio with a participation rate of 98.5 and 99.0%, respectively (average 98.8%). There were some differences in the general characteristics of the two samples, but the possibility of pooling them was shown several times in the past [23–29]. Therefore, from now on, we will deal with the lumped samples and a total of 1,712 men.

The entry examination, conducted by an international team of doctors, nurses and technicians, included a number of questionnaires, tests, diagnostic procedures, physical and chemical measurements following standard procedures described elsewhere [19, 20].

For the purpose of this analysis, only a limited number of variables was considered. They are: age (in years), rounded off at the nearest birthday; systolic blood pressure (in mm Hg), taken in supine position at the end of a physical examination (expressed as the average of two consecutive measurements), obtained following the technique reported elsewhere [30]; serum cholesterol (in mg/dl), measured on casual blood samples following the technique of Anderson and Keys [31]; cigarette smoking (in numbers of cigarettes smoked on the average per day), as elicited by a questionnaire, and body mass index (in kg/m^2), the measurement of weight and height being obtained in light undergarments, following the technique reported elsewhere [30].

The same measurements, following the same techniques, were obtained beyond year 0, also at years 5 and 10 of follow-up, when the participation rates among the survivors were 96.4 and 87.4%, respectively.

For the purpose of this analysis, men who had already suffered a major coronary event at the time of examination at year 0, 5 or 10 were excluded from the analyses starting at each of those times, as well as those with missing information of even one of the risk factors mentioned above.

The exclusion of CHD-prevalent cases at year 0 was based on the criteria corresponding to CHD prevalence codes 00–04 as adopted by the Seven Countries Study Group [2, 20] (a combination of history and ECG data).

The end point of the analysis was the first major coronary event (either fatal or nonfatal) occurring during defined periods of time in those originally free of them. They were identified by a complex monitoring system including: (a) periodical field examinations carried out at years 5, 10, 20 and 25 of follow-up; (b) periodical information on deaths and on their causes, obtained from the local register office; (c) periodical visits to local doctors and hospitals with interviews and reviews of clinical records, and (d) interviews with patients and their relatives, and the relatives of dead persons.

Such a complex and expensive system gave reasonable guarantees of completeness [32, 33].

The first major coronary event included one of the following manifestations: (1) sudden coronary death (within 2 h from onset of symptoms); (2) definite fatal nonsudden myocardial infarction (from 2 h to 30 days after the onset of symptoms); (3) nonsudden coronary death (by arrhythmias or heart failure), and (4) definite nonfatal myocardial infarction (surving more than 30 days).

Details on the diagnostic criteria were reported in other papers [32, 33]. Each event was associated with the date of occurrence.

The analysis was based on the use of the proportional hazards Cox model [34], where the fatal non-CHD deaths were treated as censored cases.

Several solutions were solved using different entry examinations, different lengths and periods of follow-up, and different sets of risk factors. In particular, in some solutions the changes of risk factors in the first 10 years were added to the risk factors measured at entry

Table 1. Structure of eight solutions of the Cox model predicting the first major CHD event as a function of 5 risk factors

Factors at year	Events in the years	Numer-ator	Denom-inator
0	0–10 (10)	81	1,624
0	0–25 (25)	288	1,624
0	0–15 (15)	136	1,624
5	5–20 (15)	172	1,460
10	10–25 (15)	196	1,336
0	10–25 (15)	207	1,389
0+DIFF (10–0)	10–25 (15)	196	1,336
0+DELTA (10–5–0)	10–25 (15)	196	1,336

Table 2. Mean levels of the selected risk factors in men free from major CHD events at year 0, 5 and 10

Factors	Mean levels		
	year 0 n = 1,624	year 5 n = 1,460	year 10 n = 1,336
Age, years	49.7 ± 5.1	54.6 ± 5.0	59.4 ± 5.0
Body mass index, kg/m^2	25.1 ± 3.7	25.7 ± 3.8	26.0 ± 4.0
Cigarettes, n/day	8.7 ± 9.9	8.5 ± 9.9	6.0 ± 8.3
Systolic blood pressure, mm Hg	143.3 ± 20.8	148.6 ± 22.5	153.1 ± 23.0
Serum cholesterol, mg/dl	201.6 ± 41.1	215.6 ± 41.4	221.7 ± 48.0

Table 3. Solutions of the Cox model predicting CHD events in 10 and 25 years, in 1,624 men aged 40–59 with risk factors measured at year 0

Factors	A		B	
	coefficient	t value	coefficient	t value
Age, years	0.0416	1.79	0.0656	5.29
Body mass index, kg/m^2	–0.0055	–0.18	0.0128	0.76
Cigarettes, n/day	0.0045	0.40	0.0128	2.20
Systolic blood pressure, mm Hg	0.0150	3.04	0.0147	5.13
Serum cholesterol, mg/dl	0.0072	2.73	0.0059	4.18

Solution A: 81 events occurring in 10 years (from year 0 to year 10); solution B: 288 events occurring in 25 years (from year 0 to year 25).

Table 4. Solution of Cox models predicting CHD in 15 years in men of different ages

Factors	A		B		C	
	coefficient	t value	coefficient	t value	coefficient	t value
Age, years	0.0602	3.35	0.0578	3.52	0.0754	4.96
Body mass index, kg/m^2	0.0004	0.02	0.0191	0.94	0.0299	1.62
Cigarettes, n/day	0.0159	1.94	0.0103	1.33	0.0319	3.95
Systolic blood pressure, mm Hg	0.0143	3.63	0.0144	4.50	0.0176	5.81
Serum cholesterol, mg/dl	0.0080	3.95	0.0057	2.97	0.0033	2.18

Solution A: 136 events in 1,624 men aged 40–59; solution B: 172 events in 1,460 men aged 45–64; solution C: 196 events in 1,336 men aged 50–69.

examination, with the purpose of testing their ability of contributing to the prediction of CHD events occurring after the first 10 years of follow-up.

Two different indicators were employed for defining risk factor changes, i.e. (a) the difference in levels of year 10 minus levels of year 0 (DIFF), and (b) the time-factor integral computed on the measurements made at years 0, 5 and 10 (DELTA). The latter indicator of changes corresponds to the mean level of the factor above or below the entry level to which men were exposed times the 10 years of exposure. Details on such procedure were presented and already employed elsewhere [17, 18].

The different solutions yielded different denominators (men exposed to the risk of the first major CHD event) and numerators (events) whose size is reported analytically in table 1.

Table 5. Solution of a Cox model predicting 207 CHD events in 15 years between years 10 and 25, in 1,389 men originally aged 40–59, as a function of risk factors measured at year 0

Factors	Solution for events	
	coefficient	t value
Age, years	0.0748	5.13
Body mass index, kg /m^2	0.0199	0.98
Cigarettes, n/day	0.0160	2.35
Systolic blood pressure, mm Hg	0.0147	4.17
Serum cholesterol, mg/dl	0.0054	3.23

Results

Table 2 reports the mean levels of the selected risk factors at years 0, 5 and 10, showing the time trends of the factors.

Solutions reported in table 3 refer to factors measured at year 0 and to follow-ups of 10 and 25 years. In both cases, systolic blood pressure and serum cholesterol are significantly predictive of CHD events. The coefficients of age and smoking habits do not reach the critical level of $p < 0.05$ in the solution at year 10, becoming significant for the 25-year follow-up solution. Body mass index never carries significant coefficients.

The predictive power of the risk factors is tested, in the solutions of table 4, for measurements taken at year 0 (men aged 40–59), at year 5 (men aged 45–64) and at year 10 (men aged 50–69), all with a uniform follow-up of 15 years. The picture is rather similar to that of the previous models with all the factors, except body mass index, producing significant coefficients. However, the coefficient of cigarette smoking does not reach the critical level of $p < 0.05$ in solutions concerning men

Table 6. Solutions of Cox models predicting CHD events in 1,336 men aged 40–59

Factors	Mean levels	Solution with DIFF		Solution with DELTA	
		coefficient	t value	coefficient	t value
Age, years	49.3±5.0	0.0700	4.54	0.0697	4.48
Body mass index, kg/m²	25.2±3.6	0.0181	0.85	0.0166	0.78
Cigarettes, n/day	8.4±9.8	0.0322	3.88	0.0311	3.95
Systolic blood pressure, mm Hg	141.5±19.3	0.0214	5.52	0.0207	5.31
Serum cholesterol, mg/dl	200.7±40.1	0.0052	2.74	0.0066	3.35
DIFF					
Body mass index	0.8±2.1	0.0370	1.05	–	–
Cigarettes	–2.4±7.9	0.0283	2.54	–	–
Systolic blood pressure	11.6±19.0	0.0135	3.55	–	–
Serum cholesterol	21.0±41.6	0.0019	1.03	–	–
DELTA					
Body mass index	4.9±12.0	–	–	0.0053	0.85
Cigarettes	–6.2±46.9	–	–	0.0055	3.03
Systolic blood pressure	60.9±115.9	–	–	0.0019	3.09
Serum cholesterol	130.0±231.4	–	–	0.0008	2.29

Solution with DIFF includes factors measured at year 0 and changes in 10 years given by the difference in levels at year 10 minus levels at year 0; solution with DELTA includes factors measured at year 0 and changes in 10 years given by the integral factor-time of measurements at years 0, 5 and 10. In both cases, 196 CHD events occurring in 15 years between years 10 and 25.

Table 7. Solution of the Cox model, stepwise technique, predicting CHD events in 1,336 men aged 40–59

Factors	Order of entrance	Final coefficient	t value	Improvement, χ^2
Age, years	forced	0.0695	4.48	–
Body mass index, kg/m²	forced	0.0155	0.73	–
Cigarettes, n/day	forced	0.0307	3.90	–
Systolic blood pressure, mm Hg	forced	0.0206	5.30	–
Serum cholesterol, mg/dl	forced	0.0068	3.47	–
DELTA				
Systolic blood pressure	1	0.0019	3.24	9.40
Cigarettes	2	0.0054	2.99	8.64
Serum cholesterol	3	0.0008	2.49	6.08
Body mass index	not entered	–	–	–

Risk factors measured at year 0 (forced) and changes in 10 years given by the integral factor-time (DELTA) of measurements at years 0, 5 and 10 (free to enter); 196 events occurring in 15 years, between years 10 and 25.

aged 40–59 and 45–64, whereas it becomes significant in the solution of men aged 50–69, likely favored by a greater incidence.

The coefficients of each factor at different ages and for the same length of follow-up were compared by the t test, but no difference was found to be significant.

The risk factor levels taken at year 0 are also predictive of the events occurring between years 10 and 25 (table 5) as well as those measured at years 10 and related to the same period of follow-up already reported in table 4.

The first solution given in table 6 deals with the baseline risk factor levels plus their changes between years 0 and 10 (year 10 minus year 0 = difference, DIFF), for the prediction of events occurring between years 10 and 25. The baseline risk factor levels retain their predictive value, but, in addition, differences in 10 years of cigarette smoking and systolic blood pressure also contribute to the prediction of coronary events. On this occasion, changes in cholesterol and body mass index provide coefficients which do not reach statistically significant levels.

When a similar solution includes the changes expressed by the DELTA (where also the 5-year measurements are considered), the picture is almost the same, but serum cholesterol changes are also contributing to the prediction of events.

All this implies that increasing levels in 10 years of the three main risk factors, i.e. cigarette smoking, systolic blood pressure and serum cholesterol, are associated with an increased risk, whereas decreasing levels are associated with a decreased risk. This is true beyond and above the role of the entry levels which maintain their predictive power. Both entry level and changes in body mass index are never associated with the occurrence of CHD events.

The solution reported in table 7 was produced by the stepwise technique after having

Table 8. Estimated changes in coronary risk following defined changes in risk factors expressed in DELTA units

	DELTA	Changes in coronary incidence, %
Body mass index	–20	–11.4
Body mass index	–10	–7.0
Body mass index	0	–2.3
Body mass index	10	2.5
Body mass index	20	7.6
Cigarettes	–600	–24.0
Cigarettes	–20	–6.7
Cigarettes	0	3.2
Cigarettes	20	14.0
Cigarettes	100	68.0
Systolic blood pressure	–100	–24.7
Systolic blood pressure	–50	–17.7
Systolic blood pressure	0	–10.1
Systolic blood pressure	50	–1.9
Systolic blood pressure	100	7.0
Serum cholesterol	–200	–21.7
Serum cholesterol	–100	–15.6
Serum cholesterol	0	–9.1
Serum cholesterol	100	–2.2
Serum cholesterol	200	5.3
Body mass index	–20	
Cigarettes	–60	–61.1
Systolic blood pressure	–100	
Serum cholesterol	200	
Body mass index	–10	
Cigarettes	–20	–40.2
Systolic blood pressure	–50	
Serum cholesterol	–100	
Body mass index	0	
Cigarettes	0	–17.8
Systolic blood pressure	0	
Serum cholesterol	0	
Body mass index	10	
Cigarettes	20	12.2
Systolic blood pressure	50	
Serum cholesterol	100	
Body mass index	20	
Cigarettes	100	100.5
Systolic blood pressure	100	
Serum cholesterol	200	

forced the baseline levels of risk factors. The improvement χ^2 suggests the progressive significant contribution of the DELTAs in improving the goodness of prediction, with the entrance of DELTA-systolic blood pressure, DELTA-cigarettes and DELTA-cholesterol in order.

If the coefficients of the solution including the DELTAs are applied back to arbitrary levels of entry values and of changes, some interesting information can be gathered (table 8). First of all, the risk of a man with baseline risk factor levels equal to their means and with changes in 10 years which follow the 'natural trend' is higher (by 17.8%) than the risk of a man with the same baseline levels and no changes at all (0.1541 vs. 0.1267). Moreover, starting again from a baseline corresponding to the mean risk factors levels, arbitrary DELTAs, above or below the 'natural trend', are associated with changes in coronary risk, in the same direction, as given by the examples of table 8. It is clear that relatively small changes in risk factors, mainly if combined together, can yield large differences in coronary incidence. As an example, in 10 years, an average reduction of 2 units of body mass index, of 6 cigarettes/day, of 10 mm Hg of systolic blood pressure and of 20 mg/dl of serum cholesterol is associated with a reduction of 61.1% in coronary incidence.

Discussion

This analysis confirms the predictive power of the three major coronary risk factors in an Italian male population sample. Their role was already known from previous analyses [20, 22, 23–29], which were in agreement with some of the major studies conducted elsewhere [1, 3–7].

This peculiar analysis suggests that the first major CHD event is predicted with similar power by the three major risk factors in men aged 40–49, 45–64 and 50–59 with a follow-up of 15 years, and in men aged 40–59 with a follow-up of 10 and 25 years. The baseline measurements are also positively associated with events occurring between years 10–25 of follow-up, but this kind of prediction can be substantially improved by adding some indicators of the changes in risk factors occurring during the first 10 years of observation in the model.

Before this paper, this approach, using the same material, was used only for systolic blood pressure [17, 18]. The models solved now indicate that the changes in cigarette smoking and serum cholesterol also contribute in improving the prediction of CHD events. In essence, the 'spontaneous' changes, whatever might be their origin, represent a kind of natural experiment where people, whose risk factor levels have changed in different amounts and directions, are compared to each other along a continuous scale of changes.

In other papers, we already emphasized the interest of this approach when related to blood pressure. In fact, it was shown that relative changes in blood pressure, mainly in the area of levels which are not hypertensive, were associated to an excess or to a reduction in risk depending upon the direction of the changes. Similar conclusions are reached now for the other two main factors. Our conclusions are in line with those observed by others on the relationships of risk factor changes with CHD incidence-mortality although obtained through different approaches [8–16].

It is clear that we are mainly dealing with relative changes whose meaning is somewhat different from factor to factor.

In fact, from table 2, it appears that cigarette smoking is a universally decreasing habit with aging, and therefore the differences in changes among individuals are mainly of relative nature.

Blood pressure, even starting from middle-aged subjects, is increasing on the average with aging, and therefore the beneficial relative changes are mainly those corresponding to a 'brake' to the increasing trends.

For serum cholesterol, the situation is more complex. It is known that in many male populations the mean levels do not increase after the age of 50–55 years. However, this 'physiological' trend is disturbed in this sample by the presence of men aged less than 50 years at entry examination and by an overall increasing secular trend which occurred in Italy and also in this samples in the 60s, i.e. the decade which was considered for the computation of changes. As well as for blood pressure, the beneficial changes are likely those corresponding not only to a decrease but also to a limited increase in cholesterol levels.

We do not know why only some people showed increased or decreased – at least relatively – blood pressure or cholesterol levels. Treatment of hypertensives was rare in the 60s [17], and dietary attention in relation to blood lipids was still an unknown practice mainly in the rural communities where the population samples were drawn.

Part of the changes of risk factor levels could be attributed to the regression toward the mean, but this phenomenon should have been largely overwhelmed by the long time trend corresponding to an observation of 10 years.

In conclusion, the long-term prediction of major CHD events cannot neglect the role of risk factor changes along time since they add power to the predictive models. In general and independently from the baseline levels, this also suggests that the lowering of major risk factors or their limited increase during aging represents a beneficial component in limiting the occurrence of the disease. If these changes are obtained through physiological means, they support the theory and the practice of primary prevention without fears of undesirable side effects.

References

1 Dawber TR: The Framingham Study. The Epidemiology of Atherosclerotic Disease. Cambridge, Harvard University Press, 1980, pp 1–255.

2 Keys A, Aravanis C, Blackburn H, Buzina R, Djordjevic BS, Dontas AS, Fidanza F, Karvonen MJ, Kimura N, Menotti A, Mohacek I, Nedeljkovic S, Puddu V, Punsar S, Taylor HL, van Buchem FSP: A Multivariate Analysis of Death and Coronary Heart Disease. Cambridge, Harvard University Press, 1980, pp 1–381.

3 Pooling Project Research Group: Relationship of blood pressure, serum cholesterol, smoking habits, relative weight and ECG abnormalities to incidence of major coronary events. Final report of the Pooling Project. J Chronic Dis 1978;31: 201–306.

4 Report of Inter-Society Commission for Heart Disease Resources: Optimal resources for primary prevention of atherosclerotic diseases. Circulation 1984;70:55A–205A.

5 Kannel WB, Keaton JD, Wentworth D, Thomas HE, Stamler J, Hulley SB, Kjelsberg MO (for the MRFIT Research Group): Overall and coronary heart disease (CHD) mortality in relation to major risk factors in 325,348 men screened for the MRFIT. Am Heart J 1986;112:825–836.

6 Martin MJ, Hulley SB, Browner WS, Kuller LH, Wentworth D: Serum cholesterol, blood pressure and mortality; Implications from a cohort of 361,662 men. Lancet 1986; ii:933–936.

7 ERICA Research Group: Prediction of coronary heart disease in Europe. The 2nd report of the WHO-ERICA Project. Eur Heart J 1991;12:291–297.

8 Hofman A: Change viewed on the level. Int J Epidemiol 1983;12:391–392.

9 Hofman A, Feinleib M, Garrison R, Van Laar A: Does change in blood pressure predict heart disease? Br Med J 1983;286:267–279.

10 Truett J, Sorlie P: Changes in successive measurements and the development of disease. The Framingham Study. J Chronic Dis 1971;24:349–361.
11 Hsu PH, Mathewson FAL, Abu-Zeid HAH, Rabkin SW: Change in risk factors and the development of chronic disease. A methodological illustration. J Chronic Dis 1977;30: 567–584.
12 Farchi G, Capocaccia R, Verdecchia A, Menotti A, Keys A: Risk factor changes and coronary heart disease in an observational study. Int J Epidemiol 1981;10:31–40.
13 Woodbury MA, Manton KG, Stallard E: Longitudinal analysis of the dynamics and risk of coronary heart disease in the Framingham Study. Biometrics 1979;35:575–585.
14 Kahn HA, Dawber TR: The development of coronary heart disease in relation to sequential biennial measurements of cholesterol in the Framingham Study. J Chronic Dis 1966;19:611–620.
15 Woodbury MA, Manton KG, Stallard E: A dynamic analysis of the chronic disease development: A study of sex-specific changes in coronary heart disease incidence and risk factors in Framingham. Int J Epidemiol 1981;10:355–366.
16 Wu M, Ware JH: On the use of repeated measurements in regression analysis with dichotomous responses. Biometrics 1979;35:513–521.
17 Menotti A, Seccareccia F and the Italian Research Group of the Seven Countries Study: Spontaneous changes of systolic blood pressure as predictors of future fatal events. Am J Hypertens 1990;3:549–554.
18 Menotti A, Keys A, Blackburn H, Karvonen MJ, Punsar S, Nissinen A, Pekkanen J, Kromhout D, Giampaoli S, Seccareccia F, Fidanza F, Nedeljkovic S, Aravanis C, Dontas A, Toshima H: Blood pressure changes as predictors of future mortality in the Seven Countries Study. J Hum Hypertens 1991;5:137–144.
19 Keys A, Aravanis C, Blackburn H, Van Buchem FSP, Buzina R, Djordjevic BS, Dontas AS, Fidanza F, Karvonen MJ, Kimura N, Lekos D, Monti M, Puddu V, Taylor HL: Epidemiological studies related to coronary heart disease: Characteristics of men aged 40–59 in Seven Countries. Acta Med Scand 1967;180:1–392.
20 Keys A: Coronary heart disease in Seven Countries. Circulation 1970; 41:1–211.
21 Keys A, Menotti A, Aravanis C, Blackburn H, Djordjevic BS, Buzina R, Dontas A, Fidanza F, Karvonen MJ, Kimura N, Mohacek I, Nedeljkovic S, Puddu V, Punsar S, Taylor HL, Conti S, Kromhout D, Toshima H: The Seven Countries Study. 2289 deaths in 15 years. Prev Med 1984; 13:141–154.
22 Menotti A, Keys A, Aravanis C, Blackburn H, Dontas A, Fidanza F, Karvonen MJ, Kromhout D, Nedeljkovic S, Nissinen A, Pekkanen J, Punsar S, Seccareccia F, Toshima H: Seven Countries Study. First 20 years mortality data in 12 cohorts of six countries. Ann Med 1989;21: 175–179.
23 Puddu V, Menotti A: An Italian study on epidemiology of ischaemic heart disease. Acta Cardiol 1969;24: 557–578.
24 Menotti A, Puddu V: Epidemiology of coronary heart disease. A 10 years study of two Italian population groups. Acta Cardiol 1972;28:66–68.
25 Menotti A, Conti S, Giampaoli S, Mariotti S, Signoretti P: Coronary risk factors predicting coronary and other causes of death in fifteen years. Acta Cardiol 1980;35:107–120.
26 Italian Research Group of the Seven Countries Study: Incidence and prediction of coronary heart disease in two Italian rural population samples followed-up for 20 years. Acta Cardiol 1982;37:129–145.
27 Italian Research Group of the Seven Countries Study: Twenty-five year incidence and prediction of coronary heart disease in two Italian rural population samples. Acta Cardiol 1986;41:283–299.
28 Menotti A, Mariotti S, Seccareccia F, Giampaoli S: The 25-year estimated probability of death for some specific causes as a function of twelve risk factors in middle-aged men. Eur J Epidemiol 1988;4:60–67.
29 Menotti A, Keys A, Kromhout D, Nissinen A, Blackburn H, Fidanza F, Giampaoli S, Karvonen MJ, Pekkanen J, Punsar S, Seccareccia F: Twenty-five year mortality from coronary heart disease and its prediction in five cohorts of middle-aged men in Finland, the Netherlands and Italy. Prev Med 1990;19: 270–278.
30 Rose GA, Blackburn H: Cardiovascular survey methods, ed 1. Geneva, WHO, 1968.
31 Anderson JT, Keys A: Cholesterol in serum and lipoprotein fractions: Its measurement and stability. Clin Chem 1956;2:145–159.
32 Menotti A, Verdecchia A, Dima F: The estimate of coronary incidence following different case finding procedures. Eur Heart J 1989;10:562–572.
33 Menotti A, Giampaoli S, Seccareccia F, Spagnolo A, con la collaborazione tecnica di Giuli B, Di Carlo G, Dima F, Pasquali M: La predizione multivariata del primo evento coronarico maggiore. Un'esperienza italiana su 25 anni di follow-up. Recenti Prog Med 1990;81:238–244.
34 Cox DR: Regression models and life tables. J R Stat Soc 1972;B43:187–220.

Cardiology 1993;82:181–190

Juha Pekkanen[a]
Markku Tervahauta[b]
Aulikki Nissinen[b]
Martti J. Karvonen[c]

[a] Department of Environmental Epidemiology, National Public Health Institute, Kuopio, Finland;
[b] Department of Community Health and General Practice, University of Kuopio, Finland;
[c] Pioppi, Italy

Does the Predictive Value of Baseline Coronary Risk Factors Change over a 30-Year Follow-Up?

Key Words
Cohort
Long-term follow-up
Blood pressure
Cholesterol
Smoking
Obesity

Abstract
The association of baseline serum total cholesterol, systolic blood pressure, smoking and body mass index with coronary heart disease (CHD) mortality was analyzed among 1,619 men aged 40–59 at baseline. Analyses were made separately for the first, second and third decade of follow-up. Serum cholesterol and smoking more than 9 cigarettes daily were strong predictors of risk of CHD death (n = 450) occurring early and late during the 30-year follow-up. After 20 years of follow-up, systolic blood pressure was no longer associated with CHD risk. In contrast, highest tertile of body mass index (over 24.7 kg/m^2) was only then associated with increased CHD risk. The correlations between the baseline and the 30-year risk factor values were 0.42 for serum cholesterol (n = 444), 0.28 for systolic blood pressure (n = 444) and 0.57 for body mass index (n = 429). Our results showed large differences in the long-term predictive power of the classical coronary risk factors. The reasons for these differences are discussed.

Introduction

Raised serum cholesterol and blood pressure levels, cigarette smoking and obesity increase the risk of coronary heart disease (CHD), also on extended follow-ups [1–8]. Few studies have, however, analyzed whether the association between these risk factors and CHD changes during the follow-up [5, 7–9], and how this relates to tracking of risk factors [9].

To address these issues, we examined the association of baseline serum cholesterol concentration, systolic blood pressure level,

Dr. Juha Pekkanen
Department of Environmental Epidemiology
National Public Health Institute
PO Box 95
SF-70701 Kuopio (Finland)

smoking and body mass index with the risk of death from CHD during a 30-year follow-up of 1,619 rural men aged 40–59 at baseline, the Finnish cohorts of the Seven Countries Study. The study population is characterized by its high serum cholesterol concentration and high mortality from CHD [10].

Subjects and Methods

In 1959, all 1,711 men aged 40–59 living in two geographically defined rural areas in eastern and south-western Finland were invited to participate, and 1,675 (98%) were examined. Follow-up examinations were carried out after 5, 10, 15 and 30 years, and were attended by 98, 98, 96, 94 and 90% of the men still alive, respectively.

The study procedures have been described in detail previously [10–12]. Briefly, each follow-up appointment included a detailed medical examination, resting electrocardiogram and assessment of risk factor state for CHD. Blood pressure was measured twice to the nearest 2 mm Hg with a mercury manometer with the subject in supine position. Means of the two readings are used in this report. Serum total cholesterol concentration was analyzed in 1959, 1964, 1969 and 1974, by the method of Abell et al. [13] as modified by Anderson and Keys [14], and in 1984 and in 1989 with an automatic analyzer [11]. Smoking was assessed by personal interview using the same standard questionnaire [10]. Standing height was recorded to the nearest centimeter, without shoes. Body weight was measured to the nearest 0.5 kg with the men wearing light underwear. Body mass index was calculated as weight (kg) divided by height (m) squared. At the 30-year follow-up in 1989, the vital status of all the men was known. Death certificates were obtained in all cases and coded according to fixed priority rules [10]. For deaths before 1974, additional information was gathered from hospital records and interviews [10].

Men with missing information on any one of the risk factors analyzed were excluded from the present analyses (n = 56). No exclusions based on prevalent disease were done as exclusions were not considered obligatory, because the end point analyzed was mortality. Exclusions based on prevalent disease only at baseline in 1959 would also have hampered comparisons between different periods, and, on the other hand, there was no follow-up examination in 1979. Analyses were, however, also repeated after excluding as well as possible men with past myocardial infarction and typical angina pectoris at the beginning of each decade of follow-up, i.e. in 1959, in 1969 and in 1979. As there were no follow-up examinations in 1979, the exclusions in 1979 were based on information of past myocardial infarction and angina pectoris in 1974, and between 1974 and 1979 on myocardial infarction and typical angina pectoris reported in 1984. The conclusions from these analyses were the same as those presented.

Analyses were first done using three separate data sets for the three 10-year follow-up periods, which were defined as the time between follow-up examinations in 1959, 1969 and 1989, and the corresponding time of the year in 1979. Values of risk factors measured at baseline, in 1959, were used in all analyses. To test the hypothesis whether estimated odds ratios are different in the three periods, the three data sets were combined, and the whole 30-year follow-up was analyzed together with two indicator variables to represent the three periods. Interactions between the continuous risk factors and the two indicator variables for periods were then tested. Analyses were first done adjusting only for age at the beginning of the decade of follow-up and area of residence, and then adjusting for other risk factors. As estimates from these two models differed very little, the latter models are primarily presented. No interactions between risk factors and area of residence were significant at the 10% level, and the results for eastern and western Finland are therefore presented together.

Tracking of risk factors was analyzed by calculating linear correlations of the risk factors measured in 1959 with those measured in 1974 and in 1989. The proportion of men remaining in the lowest and highest tertiles in 1974 and in 1989 was calculated. Tertile cutoff points for serum cholesterol were 6.10 and 7.07 mmol/l in 1974 and 5.12 and 6.15 mmol/l in 1989, for systolic blood pressure 140 and 159 mm Hg in 1974 and 146 and 165 mm Hg in 1989, and for body mass index 23.3 and 26.5 kg/m^2 in 1974 and 24.5 and 27.7 kg/m^2 in 1989.

Continuous risk factors measured at baseline in 1959 were divided into approximate tertiles. Smoking was divided into four categories: never smokers, ex-smokers, current smokers smoking only pipes or cigars or fewer than 10 cigarettes daily, and current smokers smoking more than 9 cigarettes daily. Those who had stopped smoking less than 1 year before the examination were considered as current smokers and classified according to their past smoking habit.

Results

Mean serum total cholesterol in 1959 was 6.76 mmol/l (SD: 1.35 mmol/l). The relative position according to serum cholesterol level stayed fairly stable over the whole follow-up, i.e. there was fair tracking (table 1). Tracking was as good in the lowest as in the highest tertile. Odds ratios for the highest cholesterol tertile (table 2) tended to decrease slightly during follow-up. A similar tendency was observed, when serum cholesterol was analyzed as a continuous variable (table 3). These tendencies were, however, weak. Also, the test for interaction between serum cholesterol level in 1959 and period of follow-up showed that the differences between estimated odds ratios for the 3 decades of follow-up were nonsignificant (p = 0.9).

Mean systolic blood pressure in 1959 was 144 mm Hg (SD: 21 mm Hg). Tracking of systolic blood pressure was somewhat weaker than that of serum cholesterol (table 1). Again

Table 1. Associations of continuous risk factors measured in 1959 with measurements carried out in 1974 and 1989

Risk factor measured in 1959	Year of risk factor measurement	
	1974	1989
Serum cholesterol, mmol/l		
Subjects	1,081	444
Correlation	0.53	0.42
Both times in lowest tertile, %	58	51
Both times in highest tertile, %	59	54
Systolic blood pressure, mm Hg		
Subjects	1,134	444
Correlation	0.43	0.28
Both times in lowest tertile, %	52	45
Both times in highest tertile, %	56	44
Body mass index, kg/m²		
Subjects	1,079	429
Correlation	0.74	0.57
Both times in lowest tertile, %	73	66
Both times in highest tertile, %	69	58

Table 2. Association of serum cholesterol levels measured in 1959 with CHD mortality in 1959–1969, in 1969–1979 and in 1979–1989

Serum cholesterol (mmol/l) in 1959	n[1]	CHD deaths	Multivariate adjusted[2] odds ratio (95% confidence intervals)
Follow-up from 1959 to 1969			
≤6.1	515	20	1
6.2–7.2	546	29	1.23 (0.68–2.25)
≥7.3	558	49	2.17 (1.25–3.77)
Follow-up from 1969 to 1979			
≤6.1	422	48	1
6.2–7.2	481	53	0.90 (0.59–1.38)
≥7.3	461	91	1.77 (1.20–2.61)
Follow-up from 1979 to 1989			
≤6.1	303	38	1
6.2–7.2	332	59	1.41 (0.90–2.21)
≥7.3	309	63	1.61 (1.03–2.53)

[1] Number of men at the beginning of the decade of follow-up.
[2] Adjusted for age and smoking, systolic blood pressure and body mass index in 1959.

Table 3. Association of risk factors in 1959 with risk of CHD death in different decades of follow-up and over the whole 30-year follow-up

Period of follow-up	CHD deaths	Age-adjusted odds ratio[1] (95% confidence intervals)			Multivariate adjusted[2] odds ratio[1] (95% confidence intervals)		
		serum cholesterol (1 mmol/l)	systolic blood pressure (10 mm Hg)	body mass index (3 kg/m²)	serum cholesterol (1 mmol/l)	systolic blood pressure (10 mm Hg)	body mass index (3 kg/m²)
1959–1969	98	1.32 (1.15–1.51)	1.20 (1.10–1.31)	1.11 (0.92–1.31)	1.30 (1.13–1.49)	1.18 (1.08–1.30)	1.06 (0.86–1.29)
1969–1979	192	1.29 (1.16–1.43)	1.20 (1.11–1.30)	1.01 (0.87–1.17)	1.24 (1.11–1.40)	1.19 (1.10–1.30)	0.97 (0.83–1.21)
1979–1989	160	1.26 (1.11–1.44)	1.09 (0.98–1.20)	1.21 (1.03–1.43)	1.22 (1.07–1.40)	1.05 (0.95–1.17)	1.17 (0.99–1.41)
1959–1989	450	1.29 (1.20–1.39)	1.17 (1.12–1.23)	1.10 (1.00–1.21)	1.25 (1.16–1.34)	1.16 (1.10–1.22)	1.05 (0.95–1.16)

1 Odds ratios calculated for the given change in the risk factor.
2 Adjusted for age, smoking, and other risk factors in the table.

Table 4. Association of systolic blood pressure levels measured in 1959 with CHD mortality in 1959–1969, in 1969–1979 and in 1979–1989

Systolic blood pressure (mm Hg) in 1959	n[1]	CHD deaths	Multivariate adjusted[2] odds ratio (95% confidence intervals)
Follow-up from 1959 to 1969			
≤133	534	19	1
134–149	554	26	0.96 (0.52–1.80)
≥150	531	53	1.67 (0.94–2.98)
Follow-up from 1969 to 1979			
≤133	473	42	1
134–149	485	61	1.37 (0.89–2.10)
≥150	406	89	2.38 (1.56–3.63)
Follow-up from 1979 to 1989			
≤133	363	52	1
134–149	360	65	1.18 (0.78–1.78)
≥150	221	43	1.15 (0.71–1.85)

1 Number of men at the beginning of the decade of follow-up.
2 Adjusted for age and smoking, serum cholesterol and body mass index in 1959.

Table 5. Comparison of smoking status in 1959 with that in 1974 and in 1989

Smoking status in 1959		Smoking status in 1974			Smoking status in 1989		
		never and ex-smokers	current smokers	all men	never and ex-smokers	current smokers	all men
Never smokers	n	219	4	223	110	1	111
	%	98	2	100	99	1	100
Ex-smokers	n	172	15	187	84	1	85
	%	92	8	100	99	1	100
Current smokers							
≤9 cigarettes/day	n	92	72	164	60	11	71
	%	56	44	100	85	15	100
>9 cigarettes/day	n	176	328	504	122	52	174
	%	35	65	100	70	30	100
All men	n	659	419	1,078	376	65	441
	%	61	39	100	85	15	100

there was no difference in tracking of the high or low blood pressure values. The highest systolic blood pressure tertile was less strongly associated with CHD risk during the 3rd decade of follow-up (table 4). Only 42% of the men in the highest blood pressure tertile were still alive after 20 years of follow-up, as compared to 51% of the heavier smokers and 55 and 59% of the men in the highest cholesterol and body mass index tertiles, respectively. The respective figures at the 30-year follow-up were 17, 30 and 33% for the highest systolic blood pressure, serum cholesterol and body mass index tertiles and 24% for the heavier smokers. When systolic blood pressure was analyzed as a continuous variable, it was significantly associated with CHD risk only in the 1st and 2nd decade of follow-up (table 3). The test for interaction between systolic blood pressure level in 1959 and period of follow-up showed that the decline in odds ratio after 20 years of follow-up was marginally significant ($p = 0.07$).

In 1959, 18% of the men were never smokers, 16% ex-smokers and 66% current smokers. It was very uncommon for never smokers or ex-smokers in 1959 to start smoking during follow-up (table 5). On the other hand, especially current smokers smoking less than 10 cigarettes daily at baseline tended to quit smoking during follow-up. Both ex-smokers and men smoking more than 9 cigarettes daily in 1959 were at increased risk for CHD during the entire follow-up period (table 6). The large odds ratios observed during the 1st decade of follow-up are unstable due to the very small number of CHD deaths among never smokers. There was a suggestion for the odds ratio of lighter smoking to decline during the follow-up. The test for interaction between smoking status in 1959 and decade of follow-up was, however, nonsignificant ($p = 0.25$).

Mean body mass index in 1959 was 23.7 kg/m^2 (SD: 3.2 kg/m^2). Body mass index had better tracking than both serum cholesterol and systolic blood pressure (table 1). The

Table 6. Association of smoking in 1959 with CHD mortality in 1959–1969, in 1969–1979 and in 1979–1989 and in the whole 30-year follow-up

Smoking in 1959	n[1]	CHD deaths	Multivariate adjusted[2] odds ratio (95% confidence intervals)
Follow-up from 1959 to 1969			
Never smokers	293	5	1
Ex-smokers	259	19	3.76 (1.36–10.39)
Current smokers			
≤9 cigarettes/day	228	12	2.85 (0.97–8.41)
>9 cigarettes/day	839	62	4.23 (1.65–10.84)
Follow-up from 1969 to 1979			
Never smokers	268	31	1
Ex-smokers	219	33	1.29 (0.75–2.21)
Current smokers			
≤9 cigarettes/day	196	18	0.76 (0.41–1.43)
>9 cigarettes/day	681	110	1.52 (0.97–2.38)
Follow-up from 1979 to 1989			
Never smokers	201	28	1
Ex-smokers	163	31	1.39 (0.79–2.45)
Current smokers			
≤9 cigarettes/day	151	20	0.96 (0.51–1.80)
>9 cigarettes/day	429	81	1.49 (0.91–2.44)
Follow-up from 1959 to 1989			
Never smokers	293	64	1
Ex-smokers	259	83	1.55 (1.08–2.21)
Current smokers			
≤9 cigarettes/day	228	50	1.03 (0.69–1.53)
>9 cigarettes/day	839	253	1.75 (1.29–2.38)

1 Number of men at the beginning of the follow-up.
2 Adjusted for age and serum cholesterol, systolic blood pressure and body mass index in 1959.

highest body mass index tertile was associated with increased CHD risk mainly during the 3rd decade of follow-up (table 7), whereas the middle tertile of body mass index had little association with coronary risk. When analyzed as a continuous variable, body mass index was significantly associated with CHD risk only during the last decade of follow-up (table 3). The test for interaction between body mass index in 1959 and decade of follow-up was, however, non-significant ($p = 0.25$).

Discussion

The association of risk factor levels measured at middle age and 30-year risk of CHD death was studied using follow-up data of 1,619 middle-aged rural Finnish men. The predictive power of baseline level of body mass index tended to increase and of systolic blood pressure to decline during the follow-up, whereas serum cholesterol level and heavy smoking at baseline predicted coronary events occurring both early and late during

Table 7. Association of body mass index level in 1959 with CHD mortality in 1959–1969, in 1969–1979 and in 1979–1989

Body mass index (kg/m^2) in 1959	n[1]	CHD deaths	Multivariate adjusted[2] odds ratio (95% confidence intervals)
Follow-up from 1959 to 1969			
≤22.0	530	35	1
22.1–24.7	556	29	0.82 (0.49–1.39)
≥24.8	533	34	1.07 (0.63–1.82)
Follow-up from 1969 to 1979			
≤22.0	40	57	1
22.1–24.7	483	70	1.10 (0.74–1.62)
≥24.8	461	65	1.05 (0.70–1.58)
Follow-up from 1979 to 1989			
≤22.0	271	40	1
22.1–24.7	345	53	0.96 (0.61–1.52)
≥24.8	328	67	1.48 (0.94–2.33)

1 Number of men at the beginning of the decade of follow-up.
2 Adjusted for age and smoking, serum cholesterol and systolic blood pressure in 1959.

the follow-up. Few previous analyses have addressed the questions raised in the present study [5–9], and only one study [9] had information on tracking of the risk factors over time.

When baseline measurements of risk factors are used to predict events over a long follow-up period, the assumption is that at least the ranking of the subjects according to the risk factor stays the same during the follow-up, i.e. there is good tracking. This assumption is never exactly met. Both men with high and low risk factor values tend to drift towards the population mean value [15]. Also, life-style changes cause variation in the subjects' risk factor status. These variations usually tend to dilute the observed association between baseline risk factor level and coronary risk.

In our study population, as in Finland in general, considerable changes occurred in the absolute levels of the risk factors studied during the follow-up period [12]. These risk factors showed, however, fairly good tracking in the present analyses. The good tracking of the risk factors may suggest that few of the men have actively changed their life-style in a way that would have affected their risk factor status. This is understandable, as the recommendations to lower high serum cholesterol level have been introduced only lately. The experience from the present study population may, however, not apply to more recent cohorts.

Differences in tracking is probably one reason for the differences in the long-term predictive value of the risk factors studied. Body mass index had the best tracking and tended even to increase its predictive value during the follow-up, whereas systolic blood pressure had the weakest tracking and tended to lose its predictive value after 20 years of follow-up. Men smoking 9 or fewer cigarettes per day at entry quit smoking during the follow-up more often than heavier smokers. In the

25-year follow-up analyses of the present study [6], this was shown to largely explain the decline in the predictive value of baseline light smoking.

The 25-year follow-up of 829 middle-aged men in the Zutphen Study [9] found that baseline serum cholesterol was no longer significantly associated with CHD risk during the last 10 years of the follow-up. More in line with the present findings, two large cohort studies of middle-aged men with 22 [5] and 18 years of follow-up [8] have indicated that the strength of the association of baseline serum cholesterol with coronary risk changes very little during follow-up. Previous analyses of the present study population have shown that this is true also when nonfatal myocardial infarction and other signs of CHD are considered in the analyses [12]. Analyses of the Whitehall Study [7] have suggested that the predictive value of baseline serum cholesterol level may even increase with increasing follow-up. In the Zutphen Study, tracking of serum cholesterol was very similar to that observed in the present study population [9]. No data on tracking of the risk factors were reported in the other studies.

Besides good tracking, another explanation for the strong association of serum cholesterol level in middle age and coronary risk in old age would be that most of the damage caused by high serum cholesterol level would happen by middle age, and changes in later life would have little effect. However, there is good evidence that lowering serum cholesterol levels in middle age [16] reduces the risk of CHD within a couple of years, but similar evidence does not exist for the elderly. Whatever the true underlying reason, the predictive value of serum cholesterol level in middle age appears to be strikingly good up to very old age in free-living populations. The differing experiences in the Zutphen Study [9] and in the other studies [5–8] do, however, suggest that this may not be true across all populations.

In the present study, baseline systolic blood pressure was not associated with the risk of CHD death after 20 years of follow-up. This can probably not be explained by more intensive medical treatment of men with high blood pressure values, as tracking of blood pressure was similar in the lowest and in the highest blood pressure tertiles. On the other hand, selective survival may be one factor due to the very high all-cause mortality in the highest systolic blood pressure tertile. Differences in tracking are probably only a partial reason, as systolic blood pressure had only little weaker tracking than serum cholesterol, but there was a large difference in their predictive values after 20 years of follow-up. This discrepancy may be due to the fact that in the present study population systolic blood pressure measured at old age is a good predictor of coronary risk, whereas serum cholesterol measured among men aged 65 and over is no longer associated with coronary risk [17, 18]. When considering coronary risk among elderly, serum cholesterol level measured in middle age [7, 18] and blood pressure level measured in old age [17] may be the most relevant measurements.

Stopping smoking by the time of the baseline examination was similarly associated with increased coronary risk both early and late during the follow-up. Also, fewer than 10% of the ex-smokers resumed smoking during the follow-up. These findings contrast with earlier suggestions that quitting smoking reduces CHD risk within a few years [19, 20], which were also supported by earlier analyses of the present study population [6, 12]. The earlier analyses of the present study population included only men with no signs or symptoms of CHD at baseline [6, 12]. The number of never smokers in the study population is relatively small, which causes uncertainty in the estimated odds ratios. Also, many of the ex-smokers in the present study have smoked only small amounts in the past, especially during the Second World War. On the other hand, some reports have suggested that the reduction in CHD risk may not occur as rapidly after quitting smoking as has previously been thought [21]. Taken together, it appears that quitting smoking is beneficial, but more weight should be put on preventing individuals from starting smoking.

In the present study, high body mass index was significantly associated with CHD mortality only during the last 10 years of follow-up. A similar finding has been reported before [1, 22]. Analyses of the 25-year follow-up of the present study population have indicated that the predictive value of body mass index increases with longer follow-up time, especially when also nonfatal myocardial infarction is considered in the analyses [12]. One possible explanation for these findings is that low weight may reflect preexisting disease. Persons with preexisting disease have high mortality especially in the beginning of the follow-up, which can result in a stronger association between body mass index and disease later during the follow-up. Secondly, overweight has been shown to predict subsequent development of hypertension and adverse changes in lipids [23, 24]. If body mass index is a stronger risk factor for young men, as has been suggested [22, 25, 26], then longer follow-ups of the younger men might be required for the association between initial body mass index and CHD to be disclosed.

In an analysis of the present kind, it is hard to disentangle the effects of aging and period effects from the effect of the time elapsed since the risk factor in question was measured. The effect of aging is partly controlled for by the fact that the ages of the men in different periods of follow-up overlap. In the present analyses, men were 40–59, 50–69 and 60–79 years old at the start of each decade of follow-up, so the overlap is only 10 years in successive decades of follow-up. Period effects, like improvements in treatment of nonfatal coronary events, could decrease coronary mortality without affecting risk factor levels. Their effect on the estimated odds ratios in the present study is hard to evaluate, but most period effects that do not affect risk factor levels may have similar effects at all levels of the measured risk factors and thus have less effect on the estimated odds ratios.

The present analyses show that there are large differences in the long-term predictive value of the most commonly used CHD risk factors. The possibility that an increased risk associated with a given risk factor changes during the follow-up should always be considered in analyses of follow-up studies. Also, the suggestion that serum cholesterol level measured in middle age is a better predictor of coronary risk among elderly than serum cholesterol measured in old age implies that the best way to prevent CHD among the elderly may be to measure and modify cholesterol levels at younger ages.

Acknowledgements

The present study was supported by funding from the Finnish Academy of Science and from the National Institute on Aging, USA (grant number EDC-1 1 RO1 AGO8762-01A1). We are grateful to Ms. Anna-Maria Helppi for her skillful help in typing the manuscript.

References

1 Hubert HB, Feinleib M, McNamara PM, Castelli WP: Obesity as an independent risk factor for cardiovascular disease: A 26-year follow-up of participants in the Framingham Heart Study. Circulation 1983;67:968–977.

2 Keys A, Menotti A, Aravanis C, Blackburn H, Djordevic BS, Buzina R, Dontas AS, Fidanza F, Karvonen MJ, Kimura N, Mohacek I, Nedelkovic S, Puddu V, Punsar S, Taylor HL, Conti S, Kromhout D, Toshima H: The Seven Countries Study: 2289 deaths in 15 years. Prev Med 1984;13:141–154.

3 Tyroler HA, Knowles MG, Wing SB, Logue EE, Davis CE, Heiss G, Heyden S, Hames CG: Ischemic heart disease risk factors and twenty-year mortality in middle-age Evans County black males. Am Heart J 1984;108:738–746.

4 The Italian Research Group of the Seven Countries Study: Twenty-five year incidence and prediction of coronary heart disease in two Italian rural population samples. Acta Cardiol 1986;41:283–299.

5 Ragland DR, Brand RJ: Coronary heart disease mortality in the Western Collaborative Group Study. Follow-up experience of 22 years. Am J Epidemiol 1988;127:462–475.

6 Pekkanen J, Nissinen A, Puska P, Punsar S, Karvonen MJ: Risk factors and 25 year risk of coronary heart disease in a male population with a high incidence of the disease: The Finnish cohorts of the Seven Countries Study. BMJ 1989;299:81–85.

7 Shipley MJ, Pocock SJ, Marmot MG: Does plasma cholesterol concentration predict mortality from coronary heart disease in elderly people? 18 year follow-up in Whitehall Study. BMJ 1991;303:89–92.

8 Stemmermann GN, Chyou PH, Kagan A, Nomura AMY, Yano K: Serum cholesterol and mortality among Japanese-American men. The Honolulu (Hawaii) Heart Program. Arch Intern Med 1991;151:969–972.

9 Kromhout D, Bosschieter B, Drijver M, Coulander CL: Serum cholesterol and 25-year incidence of and mortality from myocardial infarction and cancer. The Zutphen study. Arch Intern Med 1988;148:1051–1055.

10 Keys A: Coronary heart disease in seven countries. Circulation 1970;41(suppl 1):1–211.

11 Nissinen A, Kivelä SL, Pekkanen J, et al: Levels of some biological risk indicators among elderly men in Finland. Age Ageing 1985;15:203–211.

12 Pekkanen J: Coronary heart disease during a 25-year follow-up. Risk factors and their secular trends in the Finnish cohorts of the Seven Countries Study; thesis. Health Services Research by the National Board of Health in Finland No 45. Helsinki, 1987.

13 Abell L, Levy B, Brodie B, Kendall F: A simplified method for the estimation of total serum cholesterol in serum and the demonstration of its specificity. J Biol Chem 1952;195:357.

14 Anderson JT, Keys A: Cholesterol in serum and lipoprotein fractions: Its measurement and stability. Clin Chem 1956;12:145–159.

15 Criqui MH, Frankville DD, Barrett-Connor E, Klauber MR, Holdbrook MJ, Turner JD: Change and correlates of change in high and low density lipoprotein cholesterol after six years: A prospective study. Am J Epidemiol 1983;118:52–59.

16 Frick MH, Elo O, Haapa K, et al: Helsinki Heart Study: Primary prevention trial with gemfibrozil in middle-aged men with dyslipidemia. Safety of treatment, changes in risk factors, and incidence of coronary heart disease. N Engl J Med 1987;317:1237–1245.

17 Nissinen A, Pekkanen J, Porath A, Punsar S, Karvonen MJ: Risk factors for cardiovascular disease among 55 to 74 year-old Finnish men: A 10-year follow-up. Ann Med 1989;21:239–240.

18 Pekkanen J, Nissinen A, Punsar S, Karvonen MJ: Cholesterol and coronary heart disease risk in elderly men: Effect of past cholesterol levels. 2nd Int Conf Prev Cardiol and 29th Annu Meet AHA Counc Epidemiol, Washington, June 18–22, 1989.

19 Hammond EC, Horn D: Smoking and death rates – Report on forty-four months of follow-up of 187,783 men. I. Total mortality. JAMA 1958;166:1159–1172.

20 Dole R, Peto R: Mortality in relation to smoking: 20 years' observations on male British doctors. Br Med J 1976;ii:1525–1536.

21 Cook DG, Pocock SJ, Shaper AG, Kussick SJ: Giving up smoking and the risk of heart attacks. A report from the British Regional Heart Study. Lancet 1986;ii:1376–1380.

22 Rabkin SW, Mathewson FAL, Hsu PH: Relation of body weight to development of ischemic disease in a cohort of young North American men after a 26 year observation period: The Manitoba Study. Am J Cardiol 1977;39:452–458.

23 Ashley FW, Kannel WB: Relation of weight change to changes in atherogenic traits: The Framingham Study. J Chronic Dis 1974;27:103–114.

24 Conti S, Szklo M, Menotti A, Pasquini P: Risk factors for definite hypertension: Cross-sectional and prospective analyses of two Italian Rural cohorts. Prev Med 1986;15:403–410.

25 The Pooling Project Research Group: The relationship of blood pressure, serum cholesterol, smoking habit, relative weight and ECG abnormalities to incidence of major coronary events: Final report of the Pooling Project. Am Heart Assoc Monogr 60. Dallas, American Heart Association, 1978.

26 Jarrett RJ, Shipley MJ, Rose G: Weight and mortality in the Whitehall Study. Br Med J 1982;285:535–537.

Cardiology 1993;82:191–222

Jeremiah Stamler[a]
Alan R. Dyer[a]
Richard B. Shekelle[b]
James Neaton[c]
Rose Stamler[a]

[a] Department of Preventive Medicine, Northwestern University Medical School, Chicago, Ill., USA;
[b] School of Public Health, University of Texas, Health Science Center at Houston, Tex., USA;
[c] Division of Biostatistics, School of Public Health, University of Minnesota, Minneapolis, Minn., USA

Relationship of Baseline Major Risk Factors to Coronary and All-Cause Mortality, and to Longevity: Findings from Long-Term Follow-Up of Chicago Cohorts

Key Words
Risk factors
Serum cholesterol
Blood pressure
Cigarettes
Diabetes
Coronay mortality
Total mortality
Longevity
Life expectancy
Prevention

Abstract

The focus here is on relationships between major risk factors and long-term mortality from coronary heart disease (CHD) and all causes, and on longevity, in Chicago cohorts: 25-year follow-up for Peoples Gas (PG) men aged 25–39 (n = 1,119), 30-year follow-up for PG men aged 40–59 (n = 1,235), 24-year follow-up for Western Electric (WE) men aged 40–55 (n = 1,882); also 15-year follow-up for five cohorts of the Chicago Heart Association (CHA) Study: men aged 25–39 (n = 7,873), 40–59 (n = 8,515), 60–74 (n = 1,490), and women aged 40–59 (n = 7,082) and 60–74 (n = 1,243); also 12-year findings for very low risk men (n = 11,098) and other men (n = 350,564) screened for the Multiple Risk Factor Intervention Trial (MRFIT). With a high degree of consistency, multivariate analyses showed independent positive relationships of baseline serum cholesterol, blood pressure and cigarette use to risk of death from CHD and all causes. For the WE cohort, with baseline nutrient data, dietary cholesterol was also independently related to these mortality risks. Combined risk factor impact was strong for both men and women of all baseline ages. Thus, for WE men, favorable compared to observed levels of serum cholesterol, blood pressure, cigarette use and dietary cholesterol were estimated to result in 24-year risk of CHD death 69% lower, all-cause death 42% lower and longevity 9 years greater. For CHA middle-aged and older women, favorable baseline levels of serum cholesterol, blood pressure and cigarette use were estimated to yield 15-year-CHD risk lower by about 60% and longevity greater by about 5 years. For MRFIT, very low risk men (serum cholesterol <182 mg/dl, systolic/diastolic blood pressure <120/<80), nonsmokers, nondiabetic, no previous heart attack), compared to all others, observed 12-year death rates were lower by 89% for CHD, 79% for stroke, 86% for all cardiovascular diseases, 30% for cancers, 21% for other causes, 53% for all causes, and longevity was estimated to be more than 9 years longer. These findings indicate great potentials for prevention of the CHD epidemic and for increased longevity with health for men and women, through improved life-styles and consequent lower risk factor levels.

Jeremiah Stamler, MD
Department of Preventive Medicine
Northwestern University Medical School
Suite 1102, 680 N. Lake Shore Drive
Chicago, IL 60611 (USA)

Introduction

This report presents Chicago findings on the long-term impact on risk of coronary and all-cause mortality, and on life expectancy, of four major risk factors. Two of the four major risk factors are components of life-style, i.e. cigarette smoking and 'rich' diet; two are biomedical traits influenced importantly by habitual eating patterns, i.e. serum cholesterol and blood pressure. The main data are from three Chicago population samples originally surveyed beginning in the late 1950s and followed prospectively since then: men employed by the Western Electric Company (WE), age 40–55 years in the fall of 1957, and men employed by the Peoples Gas Company (PG), age 40–59 on January 2, 1958, and PG men 25–39 on January 2, 1959.

In view of recent discussions as to whether long-term findings on coronary risk of men middle aged at baseline are generalizable to younger and older men, and to women, summary data are also given from the Chicago Heart Association Detection Project in Industry (CHA), for men (age 25–39, 40–59 and 60–74) and for women (age 40–59 and 60–74) when surveyed at baseline in 1967–1973. To obtain actual risk experience for persons with optimal baseline levels of major risk factors (for comparison with estimates derived from long-term follow-up of the Chicago samples), summary data are also given on low-risk and other men comprising the large cohort of 361,662 men screened in 1973–1975 in 18 US cities for recruitment for the Multiple Risk Factor Intervention Trial (MRFIT). Among all the groups studied, the MRFIT cohort was unique – because of its much greater size – in affording an opportunity to accrue actual experience of a stratum (11,098 of 361,662 men) evaluated as low risk at baseline.

Methods

Cohorts and Baseline Surveys

WE Cohort. The cohort is comprised of 2,107 men, age 40–55 in the fall of 1957, employed at the Hawthorne Works of WE in Chicago. Of 5,397 men, age 40–55, employed by the company for at least 2 years, a random sample of 3,102 was identified for possible participation in the study, and 2,080 (61.7%) agreed. They, plus 27 other men who took part in a pilot survey, make up the cohort, examined for the first time from October 1957 through December 1958, with annual examinations thereafter for 9 years for available survivors. Methods have been described in many reports [1–17]. In brief, the surveys included a medical history, complete physical examination, electrocardiogram (ECG), serum cholesterol determination by the method of Abell et al. [18] and other items. A standardized 1-hour interview based on Burke's [19] dietary history method and a questionnaire to homemakers were used at the first and second examinations to determine the kinds and quantities of foods and beverages consumed during the preceding 28 days. Plastic food models were used to assist in portion estimates, and data were obtained on food purchases and preparation by the company cafeteria. The data were analyzed at that time with a table of food composition derived from several sources [20–24] to estimate each man's usual daily intake of energy and nutrients, including cholesterol and neutral lipids. All ECGs were subsequently classified according to the Minnesota code; in this report, ECG findings are categorized as major abnormalities, minor abnormalities or no abnormalities based on the criteria of the US national cooperative Pooling Project [7]. Men with both major and minor ECG abnormalities were classified to major abnormalities.

Follow-up for the analyses of this report began with the second annual examination in order to reduce error of classification by making use of the average of first- and second-examination data for the risk factors of concern, i.e. serum cholesterol, systolic and diastolic blood pressure, cigarettes smoked per day for men smoking at the time of both examinations and nutrients. Vital status was determined for all 2,107 men on the 25th anniversary of first examination, and death certificates were obtained for all who had died. Of the 2,107 men, 62 were excluded because they did not attend the second examination. Of the remaining 2,045 men, 142 were excluded because of prevalent coronary heart disease (CHD) at baseline and/or missing values for baseline traits. Therefore, this report gives data on 1,903 WE men, age 41–57 at second examination in 1958–1959; data on baseline nutrients were obtained for 1,882 of these men.

PG Cohort: Age 40–59 on January 2, 1958. The cohort is made up of 1,594 men (age 40–59 on January 2, 1958), i.e. all men of this age employed on that date by the PG. Of these 1,594 men, 1,465 (91.9%) were examined at baseline in 1958–1959 (midpoint July 31, 1958). Methods have been described in many reports [6–9, 14, 15, 25–29]. In brief, the extensive standardized baseline survey – done as part of the company periodic medical examination, in its medical department – included a medical history, physical examination, ECG, measurement of serum cholesterol by the method of Abell et al. [18] and other procedures. Data available from the medical department included a history of alcohol abuse ('problem drinking'). ECGs were classified by the Minnesota code; as for the WE men, in analyses in this report baseline ECGs were categorized as exhibiting major, minor or no abnormalities. Vital status has been ascertained for all men of the cohort 30 years post-baseline; death certificates are available for all decedents. Of the 1,465 participants examined at baseline in 1958–1959, 1,235 were white men free of clinical CHD and with a complete data set at baseline. They constitute the cohort reported on here.

PG Cohort: Age 25–39 on January 2, 1959. The cohort is made up of 1,609 men (age 25–39 on January 2, 1959), i.e. all men of this age employed on that date by the PG. Of these 1,609 men, 1,411 (87.7%) were examined at baseline from December 1958 through December 1963 (midpoint May, 1961). The standardized examinations were done in the PG medical department, with procedures and forms corresponding to those for the older cohort. Of the 1,411 participants, 1,119 were white men free of clinical CHD and with a complete data set at baseline. They make up the cohort for this report.

CHA Cohorts. The five cohorts are age-sex-specific groups of white men and women from the overall cohort of 39,572 men and women (Asian, black, Hispanic, white, other), age 18 and older, screened at 84 cooperating companies and organizations in the Chicago area from late 1967 to early 1973. All employees in the labor forces, totaling about 75,000 people, were invited to participate, regardless of job or shift; response rate was about 53%. Several reports give the standardized methods used by the two 4-person trained teams in the CHA field effort [5, 9, 14, 15, 30–33]. In brief, the screening involved collection of demographic, medical history and medical treatment data by questionnaire; measurement of height, weight, heart rate, casual blood pressure; an ECG; collection of blood for measurement of serum total cholesterol and uric acid, and plasma glucose 1 h after a 50-gram oral glucose load. Serum cholesterol was analyzed by the AutoAnalyzer adaptation of the method of Levine and Zak [34], uric acid by the AutoAnalzyer method of Hawk et al. [35] and plasma glucose by the AutoAnalyzer adaptation of the Hoffman [36] method. The same Pooling Project classification of ECG findings was used as for the WE and PG cohorts, i.e. major abnormalities, minor abnormalities and no abnormalities. Methods of long-term follow-up for ascertainment of vital status include local procedures and use of US Social Security Administration and National Death Index resources. Death certificates have been obtained for all decedents. With mean follow-up of 15 years, vital status is known for more than 95% of men and women of the five cohorts. The five cohorts, their numbers, race-sex, baseline age and percent of the total cohort of 39,572 CHA screenees are: 7,873 white men aged 25–39 (19.9%); 8,515 white men aged 40–59 (21.5%); 1,490 white men aged 60–74 (3.8%); 7,082 white women aged 40–59 (17.9%), and 1,243 white women aged 60–74 (3.1%). Thus, together these five cohorts comprise about two thirds of all CHA screenees. (With continued follow-up during the next several years, it is anticipated that numbers of deaths will become sufficient for meaningful analyses of the cohorts of white women age 25–39, black men and black women.)

Cohort of Men Screened for the MRFIT. The cohort consists of 361,662 men, age 35–57 when screened in 1973–1975 in 18 US cities for recruitment for the MRFIT. Background and specifics of this screening effort have been described in several reports [37–43]. In brief, at 20 screening centers in 18 US cities, eligibility for the trial was assessed on the basis of each man's cardiovascular (CVD) risk factor profile; therefore, the first screening visit included measurements of blood pressure and serum cholesterol, ascertainment by questionnaire of smoking habits, including for current cigarette smokers number of cigarettes per day, and initial assessment by questionnaire of conditions for exclusion, i.e. outside the age range 35–57, drug treatment for diabetes mellitus, and previous hospitalization of 2 weeks or longer for myocardial infarction. Participation was on a voluntary basis, the most common recruitment procedure being screening at work sites and in communities. Blood pressures were measured according to a standardized protocol by trained certified staff, with participants seated. Standard mercury sphygmomanometers were used. Diastolic blood pressure was recorded as fifth Korotkoff sound. Three readings were taken on each man; the averages of the second and third values for systolic and diastolic pressure were used in the analyses reported here. Serum cholesterol was determined for each man in one of 15 standardized laboratories under the supervision of the MRFIT central laboratory in San Francisco and the Lipid Standardization Program of the Centers for Disease Control in Atlanta, Ga., USA.

Vital status of men screened for MRFIT is being ascertained on an ongoing basis through the US National Death Index. Prior to 1979, resources of the US Social Security Administration were used. With mean follow-up from baseline of 12 years, death certificates have been obtained for 94% of identified decedents. Underlying cause of death for each decedent was coded by a trained nosologist using the ninth revision of the International Classification of Diseases. Coronary deaths were defined as International Classification of Diseases codes 410–414 plus 429.9; stroke deaths, codes 430–438, and total CVD deaths, codes 390–459.

Data Analyses

For all cohorts, relationships between baseline variables and mortality from CHD and all causes were assessed with use of multivariate proportional hazards regression (Cox) analyses [44]. Relative risks (RR) – for example, for a subgroup with more favorable baseline risk factor findings, compared to a whole cohort or to a subgroup with less favorable findings – were estimated by exponentiation of Cox coefficients. Thus, to assess RR of CHD mortality associated with a habitual dietary cholesterol intake 200 mg/1,000 kcal lower in one subgroup of men than another, the equation is: $RR = e^{-(coeff \times 200)}$, where e is the base of natural (Naperian) logarithms, and coeff is the multivariate proportional hazards regression coefficient for the relationship of baseline dietary cholesterol to long-term risk of CHD death. For example, in the WE 24-year follow-up data reported in the Results, this coefficient is 0.002801. Hence, $RR = e^{-(0.002801 \times 200)} = e^{-0.5602} = 0.571$, i.e men habitually consuming dietary cholesterol at a level of 100 mg/1,000 kcal have a 42.9% lower RR [(1.000 – 0.571) × 100] than men consuming 300 mg/1,000 kcal (about 250 vs. 750 mg/day). To estimate RR for men with 200 mg/1,000 kcal higher compared to men with lower baseline intake of dietary cholesterol, the equation is: $RR = e^{(0.002801 \times 200)} = e^{0.5602} = 1.751$, or a 24-year risk of CHD death 75.1% higher.

As previously reported [16, 45, 46], for population samples with a significant independent relationship of age to long-term risk of all-cause mortality, a further use of the multivariate coefficients is to estimate impact of a baseline level – favorable or unfavorable – on life expectancy (longevity). For example, the coefficient for the relationship of dietary cholesterol to all-cause mortality in the Cox analyses of the WE data is 0.001432. For men habitually consuming dietary cholesterol at a level of 100 mg/1,000 kcal, compared to men at 300 mg/1,000 kcal, RR of all-cause mortality is: $RR = e^{-(0.001432 \times 200)} = e^{-0.2864} = 0.751$. To estimate what this implies for life expectancy, consider the coefficient for the relationship of age to all-cause mortality for this WE cohort – 0.082342 (see results, table 2). The product for exponentiation for dietary cholesterol, $-(0.001432 \times 200) = -0.2864$, is also obtained when one multiplies the coefficient for age by –3.478 $(0.082342 \times -3.478) = -0.2864$. That is, for a group of middle-aged American men, habitually consuming 200 mg/1,000 kcal less of dietary cholesterol is equivalent to being 3.48 years younger than men consuming 200 mg/1,000 kcal more of dietary cholesterol, in relation to risk of all-cause mortality. For example, this is equivalent to being age 45.52 years instead of age 49.00. From US life tables [47], life expectancy for white men aged 45.52 and 49.00 years is estimated to be 29.09 and 26.09 years, respectively. The difference between these expectations, 3.00 years, is an estimate of the greater number of years of life that can be expected by 49-year-old white American men who habitually consume 200 mg/1,000 kcal less dietary cholesterol than their contemporaries. This same procedure is applicable for estimation of impact on longevity of higher or lower serum cholesterol, higher or lower blood pressure, nonsmoking compared to a pack of cigarettes per day, etc.

In the analyses of the 12-year mortality experience for subgroups of the cohort of 361,662 men screened for eligibility for MRFIT, age-adjusted rates were computed per 10,000 person years. Direct standardization was used to adjust for differences in age distribution among the subgroups; the total group of men screened was used as the standard population.

Results

Western Electric Cohort

Baseline Descriptive Statistics. Table 1 gives baseline descriptive statistics for the WE cohort. On average, this sample of middle-aged employed men was overweight, had systolic and diastolic blood pressures above optimal (desirable) levels (i.e. above 120/80 mm Hg [48–50]), was hypercholesterolemic, had a high intake of dietary cholesterol and saturated fat, and a low intake of polyunsaturated fat, had a high prevalence rate of current smoking and among smokers a high consumption of cigarettes/day (19/day), and had a sizeable prevalence rate of men consuming more than two alcoholic drinks/day (greater than about 30 ml or 26 g/day). Thus, status with respect to life-style and life-style-related major CVD risk factors was generally unfavorable for the whole cohort; it was overall a high risk cohort.

During the 24 years for follow-up, with data on vital status known for all men in the cohort, 691 (36.3%) of the 1,903 men died (table 2). For 310 men – 16.3% of the cohort and 44.9% of all decedents – the underlying cause of death was CHD.

Relationship of Baseline Variables to 24-Year Mortality from CHD and All Causes. Table 2 gives data on the relationship of several baseline traits to 24-year risk of death from CHD and all causes. Model 1 gives coefficients and their t values from proportional hazards regression analyses without consideration of the three dietary lipid variables; model 2, with these three variables included in the regression analyses. Results are similar with the two models for the seven variables considered in both, hence results with model 2 are summarized here. Significant independent positive relationships were found with 24-year CHD mortality regressed on systolic pressure, serum cholesterol, cigarettes/day, past drinking, body mass index, major ECG abnormalities, dietary cholesterol and age. There was suggestive evidence for an independent inverse relationship of dietary polyunsaturated fat to this end point ($t = -1.766$).

With 24-year all-cause mortality as the end point, significant independent direct relationships were demonstrated for systolic pressure, serum cholesterol, cigarettes/day, past alcohol use, heart rate, major ECG abnormalities, dietary cholesterol and age (table 2). The relationship of baseline *dietary* cholesterol to 24-year risk of CHD and all-cause mortality prevailed for men with baseline serum cholesterol of <220, 220–259 and ≥260 mg/dl [17]. Dietary saturated fat and current alcohol use showed no significant independent relationship to either end point; this was also the finding for alcohol in models assessing a possible quadratic (U-shaped) relationship.

Table 1. Baseline descriptive statistics, 1,903 white men aged 41–57 in 1959, WE study

Baseline variable	Mean ± SD
Age[a], years	48.74 ± 4.40
Systolic pressure[b], mm Hg	132.34 ± 16.84
Diastolic pressure[b], mm Hg	85.84 ± 9.81
Serum cholesterol[b], mg/dl	242.07 ± 44.66
Body mass index[b], kg/m²	25.59 ± 3.16
Current smoker, no, yes (0, 1)	0.53
Past smoker, no, yes (0, 1)	0.15
Cigarettes/day[b], all men	10.20 ± 11.34
Alcohol[c], ml/day	15.90 ± 20.36
Past drinker, no, yes (0, 1)	0.05
Heart rate[b], beats/min	72.46 ± 10.56
Major ECG abnormalities[a], no, yes (0, 1)	0.04
Minor ECG abnormalities[a], no, yes (0, 1)	0.17
Dietary cholesterol[b, d] mg/1,000 kcal	236.67 ± 60.60
Saturated fat[b, d], kcal	16.41 ± 2.28
Polyunsaturated fat[b, d], % kcal	3.95 ± 0.84

[a] At second examination, 1958–1959.
[b] Based on average for each man of first and second examination measurements.
[c] First examination data, 1957–1958.
[d] Number of men with this variable is 1,882.

Table 2. Baseline variables[a] and 24-year mortality from CHD and all causes, white men aged 41–57 in 1959, WE study, proportional hazards multiple regression analyses

Variable		Model 1 (n = 1,903)		Model 2 (n = 1,882)	
		CHD deaths (n = 310)	all deaths (n = 691)	CHD deaths (n = 308)	all deaths (n = 682)
Systolic pressure mm Hg	Regression coefficient	0.017947	0.016356	0.019060	0.016010
	t value	5.255	7.081	5.649	6.834
Serum cholesterol mg/dl	Regression coefficient	0.005382	0.002390	0.005507	0.002413
	t value	4.827	2.895	4.975	2.900
Cigarettes/day	Regression coefficient	0.024738	0.029812	0.023040	0.028869
	t value	4.710	8.521	4.377	8.167
Past smoker, no, yes (0, 1)	Regression coefficient	–0.148778	0.026691	–0.103011	0.040682
	t value	–0.784	0.212	–0.541	0.320
Alcohol[b], ml/day	Regression coefficient	–0.003422	0.000545	–0.002430	0.001029
	t value	–1.132	0.299	–0.779	0.538
Past drinker, no, yes (0, 1)	Regression coefficient	0.469201	0.400834	0.430827	0.374127
	t value	2.197	2.650	2.007	2.457
Body mass index[c] kg/m^2	Regression coefficient	0.038261	0.010709	0.037974	0.008742
	t value	2.170	0.880	2.168	0.709
Heart rate, beats/min	Regression coefficient	0.006060	0.009534	0.006461	0.009615
	t value	1.090	2.568	1.168	2.567
Major ECG abnormalities, no, yes (0,1)	Regression coefficient	0.764236	0.356866	0.776370	0.384593
	t value	3.599	2.075	3.582	2.241
Minor ECG abnormalities, no, yes (0, 1)	Regression coefficient	0.139638	0.131392	0.144861	0.131234
	t value	0.937	1.336	0.972	1.328
Dietary cholesterol mg/1,000 kcal	Regression coefficient	–	–	0.002801	0.001432
	t value	–	–	2.865	2.105
Saturated fat, % kcal	Regression coefficient	–	–	0.000205	0.000068
	t value	–	–	0.733	0.359
Polyunsaturated fat % kcal	Regression coefficient	–	–	–0.001417	–0.000490
	t value	–	–	–1.766	–0.967
Age, years	Regression coefficient	0.077624	0.086205	0.051889	0.082342
	t value	5.759	9.514	6.177	8.879

[a] See table 1 for details on year(s) of baseline measurement.

[b] Similar analyses with inclusion of alcohol and alcohol squared yielded coefficients with nonsignificant t values for both end points.

[c] Similar analyses with inclusion of body mass index and body mass index squared yielded coefficients with nonsignificant t values for both end points.

Table 3. RR of 24-year mortality from CHD and all causes, and estimated greater years of life expectancy, men with favorable baseline risk factor levels compared to whole WE cohort, based on proportional hazards regression coefficients (table 2)

Variable	Favorable value	Observed value[a]	RR[b]		Greater years of life expectancy[c]
			CHD	all causes	
(1) Dietary cholesterol, mg/1,000 kcal	100	237	0.681	0.822	+2.0
(2) Serum cholesterol, mg/dl	190	242	0.751	0.882	+1.3
(3) Systolic pressure, mm Hg	118	132	0.766	0.799	+2.3
(4) Cigarettes/day	0	10	0.794	0.749	+3.0
(5) Body mass index, kg/m^2	23	26	0.892	–[d]	–[d]
(6) Heart rate, beats/min	64	72	–[d]	0.926	+0.8
Variables 1–3 combined	–	–	0.392	0.579	+5.6
Variables 1–4 combined	–	–	0.311	0.434	+8.6
Variables 1–6 combined	–	–	0.277	0.402	+9.4

a Baseline mean value for whole cohort (table 1).
b Estimated for men with favorable baseline value, compared to baseline value for whole cohort.
c Estimated for men aged 49 with baseline favorable value compared to baseline value for whole cohort.
d No significant independent relationship.

RR of 24-Year Mortality from CHD and All Causes, Men with Favorable Baseline Values for Risk Factors, Compared to Whole Cohort, and Estimated Effects on Life Expectancy. The quantitative meaning for risk of the proportional hazards regression coefficients in table 2 is not directly apparent, but – as noted in Methods – they are readily usable to estimate RR. Table 3 presents such data, with a focus on RR for men with favorable baseline values for the several traits compared to baseline values for the whole cohort. Favorable values are in accordance with judgments and recommendations in many national and international statements on coronary prevention [27, 51–59]. These estimates are deliberately presented first here, for lower-risk men compared to the whole cohort, to emphasize the quantitative potential for prevention. Thus, with habitual dietary cholesterol intake of 100 mg/1,000 kcal (about 250 mg/day), rather than the mean value of 237 mg/1,000 kcal recorded for the whole cohort, 24-year risk of CHD death was estimated to be lower by 31.9% [(1.000 – 0.681) × 100], and risk of death from all causes was estimated to be lower by 17.8%. Nonsmokers, men with favorable baseline values of serum cholesterol or of systolic blood pressure were estimated to be at substantially lower risk of both CHD and all-cause death over the long term than the whole cohort.

These estimates of RR are multiplicative. Thus for nonsmokers with low dietary cholesterol intake, the combined effect of these two more favorable life-styles is estimated as 0.681 × 0.794 for 24-year risk of CHD death, or 0.541, i.e. a risk lower by almost 50% compared to the risk for the whole cohort (rows 1 and 4, table 3). Estimates of the combined impact of dietary cholesterol, serum cholesterol, systolic blood pressure and smoking –

for men with consistently favorable values, compared to the whole cohort – for 24-year risk of CHD death are lower by 68.9% and of all death lower by 56.6% (row 8, table 3), i.e. over the years from mean age 49 to mean age 73, risk of death from CHD lower by almost 70% and from all causes lower by more than 55%. Differences are even greater when all six variables are considered together (row 9, table 3).

The last column of table 3 gives estimates of the greater number of years of life expectancy for men with favorable baseline values for the several traits, compared to men of the whole cohort. As noted in Methods, the 1979–1981 US Decennial Life Tables report that on average a 49-year-old white American man could expect 26.09 years of additional life [47]. Since the WE men were all gainfully employed, it can be assumed that their life expectancy as a group was longer than that of all age-matched white American men (the so-called 'healthy worker' phenomenon), e.g. about 29 years instead of 26. Accordingly, compared to the whole WE cohort, men with average age 49 habitually consuming 100 mg/1,000 kcal of cholesterol, with consequent estimated life expectancy greater by 2 years (table 3), could on average expect to live about an additional 31 years rather than 29 years, the expectation for the whole cohort (row 1, table 3), i.e. to age 80 on average rather than to age 78. For men aged 49 with favorable values at baseline for all four major risk factors, life expectancy was on average estimated to be greater by 8.6 years compared to that for the whole cohort (row 8, table 3), i.e. 37.6 years (to age 86.6) rather than 29 years (to age 78.0).

RR of 24-Year Mortality from CHD and All Causes, Whole WE Cohort Compared to Men with Favorable Baseline Values for Risk Factors, and Estimated Effects on Life Expectancy. As noted above, based on observed mean values for the several risk factors, the whole WE cohort was as a group at high risk. From the data in table 3, it is simple to estimate RR for the whole cohort compared to those for men with favorable baseline findings; they are the reciprocals of the RR in table 3.

RR of 24-Year Mortality from CHD and All Causes, Men with Very High Baseline Risk Factor Values Compared to the Whole WE Cohort and to Men with Favorable Values, and Estimated Effects on Life Expectancy. Impact of very adverse baseline findings can be estimated in two relevant ways, first in comparison with the whole cohort and second in comparison with men with favorable risk factor values (tables 4, 5). The first of these two comparisons is presented in table 4. For men with very high baseline values of dietary cholesterol (400 mg/1,000 kcal or about 1,000 mg/day), serum cholesterol (280 mg/dl), systolic blood pressure (162 mm Hg), cigarette use (20/day) and body mass index (30 kg/m^2), considered as single variables, RRs of CHD in 24 years were greater than for the whole cohort by about 1.6, 1.2, 1.8, 1.3 and 1.2 times, respectively. For men with very high values for the first three of these four factors (row 7, table 4), RRs were greater compared to the whole cohort by about 3.6 and 2.3 times for CHD and all-cause mortality. These data yield an estimated life expectancy shorter by 8.2 years, i.e. instead of the estimated additional 29 years to age 78 for the whole cohort, an estimated life expectancy of only about 21 years, to age 70 on average. For men with very high values for all four of the common major risk factors (row 8, table 4), RRs of 24-year death were greater than for the whole cohort by about 4.5 times for CHD and 3.1 times for all causes. These data yield an estimate of life expectancy shorter by 11.1 years, i.e. about 18 additional years, to age 67 on average, rather than the 29 additional years to age 78 estimated for the

Table 4. RR of 24-year mortality from CHD and all causes, and estimated lesser years of life expectancy, men with very high baseline risk factor values compared to whole WE cohort, based on proportional hazards regression coefficients (table 2)

Variable	Very high value	Observed value[a]	RR[b] CHD	RR[b] all causes	Lesser years of life expectancy[c]
(1) Dietary cholesterol, mg/1,000 kcal	400	237	1.579	1.263	-2.3
(2) Serum cholesterol, mg/dl	280	242	1.233	1.096	-0.9
(3) Systolic pressure, mm Hg	162	130	1.840	1.669	-5.0
(4) Cigarettes/day	20	10	1.259	1.335	-2.9
(5) Body mass index, kg/m^2	30	26	1.164	–[d]	–[d]
(6) Heart rate, beats/min	88	72	–[d]	1.166	-1.6
Variables 1–3 combined	–	–	3.582	2.310	-8.2
Variables 1–4 combined	–	–	4.510	3.084	-11.1
Variables 1–6 combined	–	–	5.250	3.596	-12.7

a Baseline mean value for whole cohort (table 1).
b Estimated for men with very high baseline value, compared to baseline value for whole cohort.
c Estimated for men aged 49 with very high baseline value compared to baseline value for whole cohort.
d No significant independent relationship.

Table 5. RR of 24-year mortality from CHD and all causes, and estimated greater years of life expectancy, men with favorable compared to men with very high baseline risk factor levels, WE cohort, based on proportional hazards regression coefficients (table 2)

Variable	Favorable value	Adverse value	RR[a] CHD	RR[a] all causes	Greater years of life expectancy[b]
(1) Dietary cholesterol, mg/1,000 kcal	100	400	0.432	0.651	+4.5
(2) Serum cholesterol, mg/dl	190	280	0.609	0.805	+2.3
(3) Systolic pressure, mm Hg	118	162	0.432	0.494	+7.6
(4) Cigarettes/day	0	20	0.631	0.561	+6.1
(5) Body mass index, kg/m^2	23	30	0.767	–[c]	–[c]
(6) Heart rate, beats/min	64	88	–[c]	0.794	+2.4
(7) Past alcohol use, no, yes (0, 1)	no	yes	0.650	0.688	+3.9
(8) Major ECG abnormality, no, yes (0, 1)	no	yes	0.460	0.681	+4.0
Variables 1–3 combined	–	–	0.114	0.259	+14.4
Variables 1–4 combined	–	–	0.072	0.145	+20.5
Variables 1–6 combined	–	–	0.055	0.109	+22.9
Variables 1–7 combined	–	–	0.036	0.075	+26.8

a Estimated for men with favorable baseline value compared to men with very high baseline value.
b Estimated for men aged 49 with favorable baseline value compared to men with very high baseline value.
c No significant independent relationship.

Table 6. Baseline descriptive statistics, PG cohorts (means ± SD)

Baseline variable	1,119 men aged 25–39 in 1959	1,235 men aged 40–59 in 1958
Age, years	33.45 ± 2.89	50.18 ± 5.67
Systolic pressure, mm Hg	125.65 ± 14.84	133.28 ± 19.28
Diastolic pressure, mm Hg	77.71 ± 8.91	81.37 ± 9.92
Serum cholesterol, mg/dl	209.35 ± 38.05	238.05 ± 46.02
Body mass index, kg/m^2	26.22 ± 3.44	26.65 ± 3.66
Current cigarette smoker, no, yes (0, 1)	0.69	0.63
Past smoker, no, yes (0, 1)	0.11	0.15
Cigarettes/day, all men	14.60 ± 11.75	13.50 ± 12.49
Problem drinker, no, yes (0, 1)	0.01	0.04
Heart rate, beats/min	70.38 ± 11.38	71.69 ± 11.52
Major ECG abnormalities, no, yes (0, 1)	0.02	0.03
Minor ECG abnormalities, no, yes (0, 1)	0.04	0.07
Education, years	11.96 ± 2.34	10.28 ± 2.66

whole cohort. Differences are even more marked for men with very high values for all six traits (last row, table 4).

RR and estimated life expectancy increments for men with favorable risk factor findings compared to men with adverse values are given for eight variables in table 5. With each variable considered singly, risk of 24-year CHD death was lower by 23–57%, risk of all-cause mortality was lower by 20–51%, and estimated years of life expectancy were greater by 2–8 years. For men with favorable levels of the first three common major risk factors combined, compared to men with very high levels of all three (row 9, table 5), RR of CHD in 24 years was 0.114, i.e. lower by 89%, and RR of all-cause death was 0.259, i.e. lower by 74%. These data yield an estimated difference in life expectancy of 14.4 years. For men with favorable values for all four of the common major risk factors, compared to men with adverse levels for all four (row 10, table 5), risks were lower for 24-year mortality from CHD and all causes by 93 and 86%, respectively. Estimated increment in additional years of life was about 20 years. This estimate is concordant with the previous ones indicating that compared to the whole cohort men with favorable findings for these four major risk factors could expect on average to have a life expectancy better by about 9 additional years, i.e. 38 more years total, to age 87 (table 3), whereas men with all four common major risk factors very high could expect on average to have a life expectancy less than that of the whole cohort by about 11 years, i.e. only 18 more years total, to age 67 (table 4), i.e. a total difference in life expectancy for these two groups of about 20 years.

As table 5 also shows, risks were also considerably lower for men without a history of past alcohol use, or without major ECG abnormalities at baseline, compared to men with such a finding. Ex-drinking of course often was a consequence of earlier alcohol abuse and a need to quit for cause. (As noted above, current alcohol use was not significantly and independently related to risk of 24-

Table 7. Baseline variables and 25-year mortality from CHD and all causes, 1,119 white men aged 25–39 on January 2, 1959, PG study, proportional hazards multiple regression analysis

Baseline variable	CHD[a]		All causes[b]	
	regression coefficient	t value	regression coefficient	t value
Serum cholesterol, mg/dl	0.011927	3.621	0.004260	1.750
Systolic pressure, mm Hg	0.016929	1.950	0.023246	4.206
Cigarettes/day	0.037362	2.916	0.027305	3.292
Body mass index (BMI)[c], kg/m^2	−0.035847	−1.297	−0.023378	−1.186
BMI squared[c]	0.000071	1.502	0.000048	1.415
Education, years	−0.043710	−0.701	−0.089292	−2.138
Age, years	−0.024837	−0.495	−0.007711	0.227

[a] Number of CHD deaths = 52.
[b] Number of all deaths = 117.
[c] Similar analyses with inclusion of BMI only yielded coefficients with nonsignificant t values for both end points.

year mortality from either CHD or all causes for the WE cohort.)

Finally, RRs for men with very high levels of the stipulated variables, compared to men with favorable values, are the reciprocals of the relative risks in table 5. For example, from row 10 of table 5: 1.000/0.072 = 13.9 times higher risk of CHD death and 1.000/0.145 = 6.9 times higher risk of death from all causes in 24 years for men with adverse values for all four common major risk factors compared to men with favorable values for all four.

Peoples Gas Cohorts

Baseline Descriptive Statistics. As shown in column 2 of table 6, baseline findings for the PG cohort of middle-aged white men surveyed in Chicago in 1958–1959 were similar to those for the WE cohort also first examined in the late 1950s. This PG cohort too was on average at high risk of CHD.

While baseline mean values of systolic and diastolic blood pressure, and serum cholesterol were lower for the younger cohort (column 1, table 6), these levels too were high, especially for men of this age, i.e. considerably above optimal for long-term CHD and CVD risk. Also, a high percentage – 69% – of these men were cigarette smokers, with a high average daily consumption of 21 cigarettes/day.

During the 25 years of average follow-up from baseline, 117 men (10.5%) of this younger cohort died, 52 (4.6%) from CHD, i.e. CHD deaths were 44.4% of all deaths (table 7).

Relationship of Baseline Variables to 25-Year Mortality from CHD and All Causes, PG Cohort Age 25–39 in 1959. Proportional hazards multiple regression analysis showed significant independent positive relationships of baseline serum cholesterol and cigarette use to CHD, and indicated a positive relationship

Table 8. Baseline variables and 30-year mortality from CHD and all causes, 1,235 white men aged 40–59 on January 2, 1958, PG study, proportional hazards multiple regression analyses

Baseline variable	CHD deaths (n = 298)		All deaths (n = 757)	
	regression coefficient	t value	regression coefficient	t value
Systolic pressure, mm Hg	0.018879	6.298	0.012901	6.502
Serum cholesterol, mg/dl	0.001148	0.910	0.001007	1.285
Cigarettes/day	0.016219	3.218	0.018899	5.974
Body mass index (BMI), kg/m^2	−0.013562	−0.885	−0.021745	−2.409
BMI squared	0.000024	0.878	0.000039	2.427
Heart rate, beats/min	0.011193	2.233	0.008762	2.688
Education, years	−0.039182	−1.727	−0.028946	−2.055
Problem drinker, no, yes (0, 1)	0.261673	0.829	0.327449	1.684
Past smoker, no, yes (0 , 1)	0.033745	0.183	0.021778	0.184
Major ECG abnormality, no, yes (0, 1)	0.398221	1.379	0.350579	1.853
Minor ECG abnormality, no, yes (0, 1)	0.288325	1.287	0.371200	2.726
Age, years	0.076122	6.644	0.089258	12.422

also for systolic pressure (t = 1.950; table 7). For the all-cause mortality end point, significant independent positive relationships were found for systolic pressure and cigarette smoking, and a significant independent inverse association for years of education. Serum cholesterol was related positively to all-cause mortality (t = 1.750). Multiple regression coefficients for the relationships of serum cholesterol and cigarette use particularly to the CHD end point were for these young adult men much larger than for the middle-aged men of the WE (table 2) and PG (table 8) cohorts.

For men with favorable baseline levels of these three major risk factors, compared to the whole of the younger PG cohort, RRs of death in 25 years – from mean age 32 to mean age 57 – were much lower (table 9). Favorable values for all three of these traits were associated with a RR for CHD death of 0.274 and for all-cause mortality of 0.387 compared to the whole cohort, i.e. risks lower by 73 and 61%, respectively (row 6, table 9). Since age was not significantly related to either CHD or all-cause mortality for this younger PG cohort, estimates could not be made – comparable to those for the middle-aged Chicago cohorts – of the greater years of life expectancy with favorable levels of these major risk factors.

Of the 1,235 PG white men aged 40–59 on January 2, 1958, 757 (61.3%) died during the 30-year follow-up period (table 8). CHD was the underlying cause of death for 298 men – 24.1% of the cohort, and 39.4% of all decedents.

Relationship of Baseline Variables to 30-Year Mortality from CHD and All Causes, PG Cohort Age 40–59 on January 2, 1958. Among the 12 variables in table 8, a significant independent positive relationship was shown with CHD mortality for systolic pressure, number of cigarettes/day, heart rate and age; education was inversely related (t = −1.727). There was a significant independent positive rela-

Table 9. RR of 25-year mortality from CHD and all causes, men with favorable baseline risk factor levels compared to whole PG young adult cohort, based on proportional hazards regression coefficients (table 7)

Variable	Favorable value	Observed value[a]	RR[b]	
			CHD	all causes
(1) Serum cholesterol, mg/dl	170	209	0.628	0.847
(2) Systolic pressure, mm Hg	110	126	0.763	0.689
(3) Cigarettes/day	0	15	0.571	0.664
(4) Education, years	16	12	–[c]	0.700
Variables 1 and 2 combined	–	–	0.479	0.584
Variables 1–3 combined	–	–	0.274	0.387
Variables 1–4 combined	–	–	–	0.271

[a] Baseline mean value for whole cohort (table 6).
[b] Estimated for men with favorable baseline value, compared to baseline values for whole cohort.
[c] t value for this end point = –0.701 (table 7).

tionship with 30-year risk of all-cause death for systolic pressure, cigarettes/day, heart rate, minor ECG abnormalities and age. A significant independent quadratic relationship was found for body mass index and all-cause mortality, and a significant independent inverse relationship for education. Problem drinking and major ECG abnormalities were positively related to risk of all-cause death ($t = 1.684$ and 1.853, respectively).

In regard to the established major risk factors for CHD, multiple regression coefficients for the middle-aged PG and WE cohorts were almost identical for systolic pressure and were similar for cigarettes/day (tables 2, 8). Serum cholesterol was significantly and independently related to risk of 24-year mortality from both CHD and all causes for the WE men (table 2) and for the younger PG men (table 7), but not to risk of 30-year mortality for the PG men aged 40–59 on January 2, 1958 (table 8).

Based on the data in table 8, RR and impact on longevity were estimated for men with favorable values for modifiable traits compared to the whole cohort (table 10). Based only on systolic pressure and smoking status, for men with favorable values at baseline, 30-year risks of death from CHD and all causes were lower by about 40 and 37% (row 5, table 10); with inclusion also of heart rate, about 45 and 41%, respectively. These data yield the estimate that for men with favorable systolic pressure and smoking status at baseline expected number of years of life was greater by about 4.3 years compared to the whole cohort; 5.0 years with favorable values for all three traits. For white men age 50, the US Decennial Life Tables for 1979–1981 show an average expectation of 25.26 years [47]. Given that the PG cohort was made up of employed white men, at average age 50 their life expectancy was somewhat greater, e.g. 28 years, to age 78. Therefore, men with favorable values for the foregoing three traits

Table 10. RR of 30-year mortality from CHD and all causes, and estimated greater years of life expectancy, men with favorable baseline risk factor levels compared to whole PG middle-aged cohort, based on proportional hazards multiple regression coefficients (table 8)

Baseline variable	Favorable value	Observed value[a]	RR[b]		Greater years of life expectancy[c]
			CHD	all causes	
(1) Systolic pressure, mm Hg	118	133	0.753	0.824	+1.8
(2) Cigarettes/day	0	14	0.797	0.768	+2.5
(3) Heart rate	64	72	0.914	0.932	+0.7
(4) Education	14	10	0.855	0.891	+1.1
Variables 1 and 2 combined	–	–	0.600	0.633	+4.3
Variables 1–4 combined	–	–	0.469	0.526	+6.1

[a] Baseline mean value for whole cohort (table 6).
[b] Estimated for men with favorable baseline value, compared to baseline for whole cohort.
[c] Estimated for men aged 50 with favorable baseline value, compared to baseline value for whole cohort.

Table 11. Major risk factors at baseline and 15-year mortality from CHD: young adult, middle-aged and older white men, middle-aged and older white women, CHA, proportional hazards multiple regression analyses

Sex-age group	Number of people	CHD deaths		Serum cholesterol		Systolic pressure		Cigarettes/day	
		total	per 1,000	coeffi-cient	t value	coeffi-cient	t value	coeffi-cient	t value
Men 25–39	7,873	57	7.2	0.0158	5.497	0.0187	2.510	0.0342	4.078
Men 40–59	8,515	458	53.8	0.0074	7.927	0.0190	9.419	0.0211	6.967
Men 60–74	1,490	209	140.3	0.0042	2.053	0.0144	4.756	0.0175	3.097
Women 40–59	7,082	123	17.4	0.0045	2.215	0.0219	5.853	0.0507	7.186
Women 60–74	1,243	90	72.4	0.0066	2.724	0.0171	4.028	0.0294	2.822

Age was included in these analyses. For the end point 15-year death from all causes, the coefficients for serum cholesterol, systolic pressure, cigarettes/day and age were: men aged 25–39: 0.006832; 0.012888; 0.028369; 0.096960; men aged 40–59: 0.002216; 0.014382; 0.026770; 0.099080; men aged 60–74: 0.000906; 0.009922; 0.023310; 0.130451; women aged 40–59: 0.000434; 0.013228; 0.029090; 0.080774, and women aged 60–74: 0.001678; 0.015018; 0.028335; 0.081279.

All coefficients were statistically significant ($p < 0.05$) except those for serum cholesterol for men aged 60–74, for women aged 40–59 and for women aged 60–74.

Table 12. RR of 15-year mortality from CHD: white men and women with favorable baseline levels of three major risk factors compared to those with observed population mean baseline levels, five cohorts, CHA, based on proportional hazards multiple regression coefficients (table 11) and estimated effects of more favorable levels on life expectancy

Cohort by sex and baseline age	Serum cholesterol	Systolic blood pressure	Cigarettes/ day	All three[a]	Estimated greater life expectancy[b] years
RR, people with favorable compared to people with observed population mean levels					
Men 25–39	0.684	0.627	0.710	0.305	+7.3
Men 40–59	0.783	0.646	0.810	0.410	+6.0
Men 60–74	0.882	0.613	0.900	0.486	+2.5
Women 40–59	0.839	0.674	0.738	0.417	+4.6
Women 60–74	0.705	0.588	0.889	0.369	+5.8
Favorable and observed population mean levels					
Men 25–39	*170* 194	*110* 135	*0* 10	–	–
Men 40–59	*180* 213	*118* 141	*0* 10	–	–
Men 60–74	*180* 210	*118* 152	*0* 6	–	–
Women 40–59	*180* 219	*118* 136	*0* 6	–	–
Women 60–74	*180* 233	*118* 149	*0* 4	–	–

[a] Estimated as the product of the three RR, e.g., for men age 25–39: 0.684 × 0.627 × 0.710 = 0.305.
[b] Based on proportional hazards multiple regression coefficients for serum cholesterol, systolic blood pressure, cigarettes/day, age.
[c] Favorable levels are in italic.

compared to the whole cohort could expect to live on average 5 + 28 = 33 additional years, i.e. to age 83.

Based on the data in tables 7 and 8, it is readily possible also to calculate RR for the two PG cohorts as a whole compared to men with favorable values of the common risk factors; they are the reciprocals of the RR in tables 9 and 10. Correspondingly, it is simple to estimate RR for PG men with favorable values compared to men with very high values for the risk factors significantly related to mortality, and vice versa, as was done for similar WE men (tables 4, 5), and for the middle-aged PG cohort to estimate comparative prognosis for longevity for such men. Space limitations preclude the display of such additional tabulations here. The results are similar to those for WE men.

Chicago Heart Association Cohorts

As noted in the Introduction, most of the long-term prospective studies on baseline traits and CHD risk have involved cohorts of men middle-aged at entry, hence questions have been posed as to the validity of generalizing their findings to younger and older men and to women. Since this matter is of importance for medical care and public health, data are included here for five cohorts from the CHA study – white men aged 25–39, 40–59 and 60–74, and white women aged 40–59 and

60–74 at baseline (tables 11, 12). For all five of these cohorts, there was a strong, graded, independent, significant positive relationship between baseline serum cholesterol, blood pressure, cigarette use and 15-year risk of mortality from CHD. For each cohort, for persons with more favorable levels of all three major risk factors compared to the whole cohort, 15-year RR of CHD was substantially lower – 70% lower for young adult men; 59% lower for middle-aged men; 51% lower for older men; 58% lower for middle-aged women; 63% lower for older women (next to last column, table 12).

As noted above for PG men aged 25–39 at baseline, for the CHA cohort aged 25–39 at entry, the relationship between serum cholesterol and CHD risk was particularly strong (tables 11, 12). A serum cholesterol level lower by 24 mg/dl was associated with a CHD risk lower by almost 32% (table 12). Coefficients for this relationship were statistically significant for all four cohorts of middle-aged and older men and women, but smaller than for young adult men, hence differences in RR – although substantial – were less than for young adult men. However, given the marked increase in CHD risk with age (table 11), estimated absolute benefits from lower serum cholesterol for the middle-aged and older cohorts over a finite number of years (e.g. 15) were substantial. Thus, for men aged 60–74 at baseline, with a mean serum cholesterol for the whole cohort of about 210 mg/dl and a 15-year CHD death rate of 140/1,000, a serum cholesterol of 180 mg/dl was associated with a RR of 0.882, i.e. a rate of 123/1,000 instead of 140, or 17 fewer CHD deaths per 1,000 persons. For comparison, for CHA men aged 25–39 at baseline, with a 7/1,000 CHD death rate, during the initial 15 years of follow-up, serum cholesterol lower by 24 mg/dl – with a RR of 0.684 – is estimated to be associated with a rate of about 5/1,000, i.e. 2 fewer CHD deaths per 1,000; however, for young adult men over the long term (e.g. by age 65 or 75), it can readily be estimated that a much larger number of CHD deaths per 1,000 will be avoided by those with favorable serum cholesterol levels if they avoid rises in those levels. Similarly, compared to the whole cohort of older women with a 15-year CHD death rate of 72/1,000, those with a serum cholesterol lower by 53 mg/dl and a consequent RR of 0.705 are estimated to experience 51 CHD deaths per 1,000, i.e. 21/1,000 fewer CHD deaths. Expected absolute benefits from favorable status in regard to blood pressure and cigarette use are even greater than the foregoing for middle-aged and older men and women.

The CHA samples of 8,515 men and 7,082 women aged 40–59 at baseline, and 1,490 men and 1,243 women aged 60–74 included enough people with a clinical diagnosis of diabetes mellitus to assess the independent impact of this condition on CHD risk. For all four cohorts, there was a strong significant positive relationship, independent of serum cholesterol, blood pressure and cigarette use. RRs were 2.35, 2.80, 3.42 and 2.32 for diabetic compared to nondiabetic subgroups from these four cohorts. For all four of those cohorts, baseline major ECG abnormalities – defined by criteria of the US national cooperative Pooling Project [7] – were also related in multivariate Cox analyses to 15-year risk of CHD death, with RRs of 3.27 and 1.84 for men; 1.47 and 1.55 for women. Also, for all four cohorts, education was inversely related to 15-year risk of CHD mortality; the more the education, the lower the risk, with relative risks in the range 0.70–0.82 with 4 additional years of education.

Cohort of Men Screened for the MRFIT

As noted in the Introduction, with the sample sizes of the WE, PG and CHA cohorts, ranging from 1,119 to 8,515 persons, and the

Table 13. Baseline risk factor levels and age-adjusted mortality per 10,000 person years, men age 35–57 screened for the MRFIT – low-risk men compared to whole cohort; to whole cohort exclusive of low-risk men; and to whole cohort exclusive of low-risk men and men with diabetes (DM) or myocardial infarction (MI)

Variable	Low-risk men (n = 11,098)		Whole cohort (n = 361,662)		Whole cohort minus low-risk men (n = 350,564)		Whole cohort minus low-risk men, men with DM, MI[a] (n = 331,717)	
Baseline risk factor levels (means ± SD)								
Serum cholesterol, mg/dl	162.4 ± 15.3		214.6 ± 39.6		216.3 ± 39.0		215.8 ± 38.5	
Blood pressure, mm Hg								
Systolic	111.0 ± 5.9		130.0 ± 15.9[b]		130.6 ± 15.7[c]		130.5 ± 15.6	
Diastolic	71.6 ± 5.5		83.9 ± 10.7		84.3 ± 10.6		84.1 ± 10.4	
Smokers, %								
Cigarettes/day	0		36.8		38.0		37.6	
All men	0.0		9.5 ± 14.9		9.8 ± 15.0		9.7 ± 14.9	
Smokers	0.0		25.9 ± 13.6		25.9 ± 13.6		25.8 ± 13.5	
Age, years	43.5 ± 6.3		46.0 ± 6.4		46.0 ± 6.4		45.9 ± 6.4	
Number of deaths and age-adjusted death rate[d]								
Mortality by cause								
CHD	22	*2.1*	8,323	*20.0*	8,301	*20.4*	6,659	*17.5*
Stroke	3	*0.4*	788	*1.9*	785	*1.9*	682	*1.8*
CVD	38	*3.8*	10,949	*26.3*	10,911	*26.9*	8,927	*23.4*
Cancer	136	*13.6*	8,061	*19.4*	7,925	*19.5*	7,403	*19.4*
Non-CVD, non-cancer	104	*9.5*	5,003	*12.0*	4,899	*12.1*	4,259	*11.1*
All	278	*26.9*	24,013	*57.6*	23,735	*58.4*	20,589	*53.9*

[a] Also excludes men with missing systolic blood pressure values.
[b] Number of men = 353,340, due to missing values.
[c] Number of men = 342,242 due to missing values.
[d] Age-adjusted death rate per 10,000 person years is in italic; mean duration of follow-up = 11.6 years.

life-styles prevailing in the United States when they were surveyed at baseline, these cohorts contained very few people with favorable levels of all the major risk factors, too few to enable meaningful measurement of their actual long-term CHD risk. Therefore, it has been possible only to *estimate* RR of CHD of such subgroups, as was done here in tables 3, 5, 9, 10, and 12. With the recent availability of data on the large cohort of 361,662 men aged 35–57 at baseline, screened in a standardized way in 18 US cities in 1973–1975 for recruitment for the MRFIT, it became possible for the first time to collect precise *observed* data on risk for men with all baseline major risk factor levels favorable (table 13). Baseline findings used for defining a low risk subgroup were: serum cholesterol <182 mg/dl (lowest quintile), systolic pressure <120 mm Hg, diastolic pressure <80 mm Hg, not a current cigarette smoker, without a history of drug treatment for diabetes

Table 14. Baseline quintiles of systolic pressure, serum cholesterol and age-adjusted CHD mortality per 10,000 person years: men free at baseline of history of hospitalization for heart attack, MRFIT primary screenees

Serum cholesterol mg/dl	Systolic pressure, mm Hg				
	<118	118–124	125–131	132–141	≥142
Nonsmokers					
<182	3.1	4.1	5.5	6.3	14.7
182–202	4.9	6.1	9.0	8.1	17.6
203–220	5.4	6.8	9.0	11.2	19.7
221–244	7.0	9.6	9.3	12.6	24.5
≥245	12.5	13.4	16.6	21.2	34.9
Smokers					
<182	11.4	10.7	15.1	15.7	30.4
182–202	10.6	13.5	18.9	21.8	35.9
203–220	15.2	16.7	21.1	29.1	44.3
221–244	19.9	23.1	23.5	33.5	57.1
≥245	26.2	31.8	36.0	42.2	64.2

Mean follow-up is 11.6 years.

mellitus, without a history of hospitalization for 2 weeks or more for a heart attack.

The first salient finding from this analysis is that only 3.1% of the cohort – 11,098 of the 361,662 men – met all six of these criteria (table 13). Men aged 35–57 with favorable status for these six traits were indeed 'rare birds' in the USA in 1973–1975. The second salient finding is their low mean levels of serum cholesterol 162 mg/dl and of blood pressure 111/72 mm Hg, which – together with 0 cigarettes/day, no history of diabetes or heart attack – do indicate truly low-risk status, from the estimates based on the data of all other prospective population studies, including those on the eight Chicago cohorts presented here. The third salient finding is the remarkably low actual observed age-standardized mortality rates during the first 11.6 years of follow-up: age-standardized CHD death rate 2.1/10,000 person years, only one tenth the rate for the rest of the cohort (columns 1 and 3, table 13); CVD death rate 3.8, 14% of the rate for the rest of the cohort; cancer and non-CVD non-cancer death rates 70 and 79% those of the rest of the cohort; mortality from all causes only 46% – less than half – that of the rest of the cohort.

For the MRFIT cohort overall, baseline serum cholesterol, systolic (or diastolic) pressure and cigarette use were strongly continuously, significantly, independently related to 11.6 year risk of death from CHD, all CVD diseases, and all causes, as were age and history of treated diabetes [39, 41–43, 48–50, 60]. These findings were shown in analyses with use of multiple regression (logistic, proportional hazards) and multiple cross-classification, e.g. 50-cell analyses with quintiles of both serum cholesterol and blood pressure for nonsmokers and smokers (table 14). For serum cholesterol, and systolic and diastolic blood pressure, risk of CHD death was progressively higher for men in the second, third, fourth and fifth quintile of baseline level, for both nonsmokers and smokers. Optimal serum cholesterol (quintile 1) was <182 mg/dl; optimal systolic pressure (quintile 1) <118 mm Hg; optimal diastolic pressure (quintile 1) <76 mm Hg. The strong graded significant relationship of serum cholesterol to CHD risk prevailed for men in every quintile of blood pressure, for both nonsmokers and smokers. The strong graded significant relationships of blood pressure (systolic, diastolic) to CHD risk prevailed for men in every quintile of serum cholesterol, for both nonsmokers and smokers. For nonsmokers free of heart attack history at baseline, age-adjusted risk of CHD death ranged from 3.1/10,000 person years for those with serum cholesterol <182 mg/dl and systolic pressure <118 mm Hg to 34.9 for those in quintile 5 of both these variables (serum cholesterol ≥245 mg/dl, systolic pressure ≥142 mm Hg), a range of over 11-fold. For smokers, this range was from 11.4 to 64.2 per 10,000 person years, i.e. almost six-fold, with consistently much higher rates for smokers than nonsmokers at any and every level of serum cholesterol and blood pressure. The overall range of CHD risk for the entire cohort – from 3.1 to 64.2 per 10,000 person years – was about 21 times, based on only a single screening visit for each man, without any correction upward for the regression dilution bias resulting from only one assessment of these traits [61]. Given the very large sample size, these estimates of risk for multiple subgroups of MRFIT men are highly precise, i.e. have narrow confidence intervals, albeit range of risk is underestimated, for the reason just cited. Qualitatively similar significant findings on the impact of these three key traits on risk of CHD death prevailed for men from each of the 20 centers in 18 cities across the USA; for men of each age group (35–39, 40–44, 45–49, 50–54 and 55–57); for men of each ethnic group (Asian, black, Hispanic, white); across socioeconomic strata; for men with and without a history of treated diabetes or of hospitalization for heart attack.

Proportional hazards (Cox) multiple regression analyses were used to estimate RR of mortality and impact on longevity with the more favorable mean levels of the major risk factors for the 11,098 low risk MRFIT men compared to the rest of the cohort (columns 1 and 3, table 13). These analyses included as regressors age, serum cholesterol, systolic pressure, cigarettes/day, history of treated diabetes (no, yes), history of hospitalization for heart attack (no, yes), race (black: no, yes), and median income of census tract of residence of each man. All of these variables were significantly and independently related to risks of mortality ($p < 0.0001$). For all-cause mortality, the coefficients were 0.002362 for serum cholesterol, 0.015113 for systolic pressure, 0.022778 for cigarettes/day and 0.087691 for age. For the 11,098 men compared to the rest of the cohort, based on the group differences in mean values at entry for serum cholesterol, systolic pressure and cigarettes/day, RR of death in 12 years was 0.541, i.e. lower by 46 %. This translated into a life expectancy estimated to be longer by 6.3 years for the group of low risk men compared to the rest of the cohort. These findings, similar to those for the Chicago cohorts, are – it is relevant to re-emphasize – underestimates because they are based on data for the three major risk factors collected at only one visit, i.e. they suffer from the effects of regression-dilution bias [61]. Moreover, another form of bias led to further underestimation of the favorable effect of low serum cholesterol on RR and longevity. This resulted from the fact that the coefficient for serum cholesterol and all-cause mortality was relatively small due to the quadratic (J-shaped) relationship between serum cholesterol and all-cause death, i.e.

higher mortality for men with very low entry serum cholesterol levels (<160 mg/dl) compared to men with higher values. Extensive evidence indicates that this association between low serum cholesterol and greater mortality is spurious, i.e. not reflective of a cause-and-effect relationship, whereas there are no data to support the notion of an etiologically significant relationship [62]. But the Cox analysis here to generate the coefficient for serum cholesterol and all-cause mortality did not correct for this problem.

To deal with this problem and the artifactual underestimation of impact on RR and longevity resulting from it, the foregoing analyses were done with all CVD deaths, rather than all deaths, as the end point. For CVD death, the coefficient for serum cholesterol was 0.005608; for systolic pressure, 0.021954; for cigarettes/day, 0.021394, and for age, 0.088774. Again, based on the differences at baseline for these three major risk factors for the 11,098 low-risk men compared to the rest of the cohort (columns 1 and 3, table 13), RR of CVD death in 12 years was 0.406, i.e. lower by 59%. This translated into a life expectancy estimated to be longer by 9.0 years for the low-risk MRFIT group compared to the rest of the cohort, based solely on the three specified major risk factors measured only once. Of these 9.0 years of estimated greater longevity, the calculations indicate that 2.9 years (32%) are attributable to group mean serum cholesterol of 162 rather than 214 mg/dl; 4.2 years (47%) to group mean systolic pressure of 111 instead of 130 mm Hg; and 1.9 years (21%) to group mean cigarettes/day of 0 instead of 9/day. Thus, favorable status in regard to each of the major risk factors contributed significantly to the estimated much greater longevity of the low-risk group compared to the rest of the cohort.

Both diabetes mellitus and heart attack history were strongly related to risk of both CVD and all-cause death in these analyses of the MRFIT data. For diabetes, the coefficient was 0.983773 for CVD death, translating into a RR 2.68 times higher for diabetic compared to nondiabetic men; 0.869693 for all causes, RR 2.39, with an estimated life expectancy shorter by 8.1 years for 46-year-old men with diabetes compared to nondiabetic men. For men with a history of hospitalization for heart attack, the coefficient for CVD death was 1.602737, RR 4.97; for all causes death, 1.133173, RR 3.11, with an estimated life expectancy shorter by 10.3 years for 46-year-old men with this history compared to men free of such a history.

For black men in the MRFIT cohort, risks of both CVD and all-cause death were greater than for all other men (predominantly white men) based on these multiple regression analyses of the 12-year mortality experience. Coefficients were 0.120770 for CVD death and 0.238660 for all-cause mortality, yielding RR estimates of 1.128 and 1.270, i.e. risk of CVD greater by 13% and of all-cause death greater by 27% for black compared to other men. This coefficient for all-cause death, together with the coefficient for age (0.087691), translated into a life expectancy shorter by 2.3 years for black compared to other men average age 46 at entry into the MRFIT cohort. The US life tables for 1979–1981 indicate that for all American black and white men aged 46, the difference in life expectancy was 3.8 years [47]. Given the more favorable socioeconomic status of the black men in MRFIT compared to all black middle-aged men, these two estimates are concordant.

Discussion

In accordance with the plan for this special number of *Cardiology,* the focus of this paper has been on long-term prognosticators of

CHD death, as well as all-cause mortality and longevity, in three Chicago cohorts with 25–30 years of follow-up from first examination. The main analyses have dealt with the combined impact of established major risk factors and the related matter of the potential for prevention and greater longevity with lifetime favorable status for the key traits, serum cholesterol, blood pressure and smoking habit (as well as dietary lipid, available for one cohort only). The findings are consistent and unequivocal for the three Chicago cohorts of employed men with 25–30 years of follow-up – men aged 40–55 in the fall of 1957, from the WE study, and men aged 40–59 on January 2, 1958 and men 25–39 on January 2, 1959, from the PG study. The validity and generalizability of the results for these three cohorts of middle-aged and young adult men are buttressed by the concordant consistent data from the CHA for five cohorts, including not only middle-aged and young adult men, but also older men aged 60–74 at baseline, middle-aged women aged 40–59 at baseline and older women aged 60–74 at baseline, all followed for 15 years after entry. They are further reinforced by the concordant consistent findings with 12-year follow-up of the uniquely large MRFIT cohort of 361,662 men aged 35–57 at baseline and its multiple subcohorts identified based on age, ethnicity, geographic location in the USA and socioeconomic status.

For the two PG and the WE cohorts with prospective data for 25–30 years from initial examination in the late 1950s or early 1960s, as well as for the five other Chicago cohorts, the whole MRFIT cohort and its subcohorts, the *main findings* were: strong, graded, independent, significant positive relationships between serum total cholesterol, blood pressure, cigarette use and long-term risk of death from CHD and all causes. These positive relationships, consistent with almost no exception, were obtained with availability of only one baseline measurement of the risk factors for most of the cohorts (two measurements for the WE cohort). Therefore, the strong effects of these etiologically significant variables on risk, even as long as 25 or 30 years after baseline, are *underestimates* due to regression-dilution bias [61].

The data permit definition for comparison of pairs of population samples based on favorable compared with usual mean levels of these three major risk factors considered together, i.e. populations with favorable levels: mean serum cholesterol 170 (young adult) or 190 (middle age) mg/dl, mean systolic pressure 110 (young adult) or 118 (middle age) mm Hg, nonsmoking compared to populations with usual observed levels for US men in the late 1950s and early 1960s: mean serum total cholesterol 209 (young adult) or 242 (middle age) mg/dl, mean systolic pressure 126 (young adult) or 132–133 (middle age) mm Hg, population mean cigarettes/day 15 (young adult) or 10–14 (middle age). For the young adult population with favorable levels for all three traits compared to the usual young adult population (PG), RR of CHD death during the ensuing 25 years was 0.274 for young adult men, i.e. lower by 73%; RR of all-cause death was 0.528, i.e. lower by 47%. For the middle-aged population with favorable levels, compared to the usual middle-aged population (PG, WE), RR of CHD death during the ensuing 30 or 24 years were 0.600 (PG) and 0.457 (WE), i.e. lower by 40 and 54%; RR of all-cause death were 0.633 and 0.528, i.e. lower by 37 and 47%.

For the WE cohort, baseline nutritional data were also available, and dietary lipid – in particular dietary cholesterol – was strongly, significantly and independently related to 24-year risk of death from CHD, all CVD and all causes [10, 12, 16, 17], a key finding concordant with results from at least three other long-term population studies (Honolulu Heart Program, Ireland-Boston Diet-Heart Study, Zutphen Study in the Netherlands) [16, 63–65]. With definition of a favorable population sample based on this variable (habitual cholesterol intake of 100 mg/1,000 kcal), as well as serum cholesterol (190 mg/dl), systolic pressure (118 mm Hg) and cigarette use (0/day), compared to the whole WE cohort with its mean levels of these traits recorded at baseline (237 mg/1,000 kcal, 242 mg/dl, 132 mm Hg, 10/day), RR of death in the ensuing 24 years (from mean age 49 to 73) was 0.311 for CHD, i.e. 69% lower, and 0.434 for all causes, i.e. 56% lower. These lower risks for the sample with favorable values for all four of the major risk factors, compared to those observed for the whole WE cohort at baseline, translate into an estimated life expectancy longer by 8.6 years – attributable to favorable status in regard to each of the four risk factors (2.0, 1.3, 2.3 and 3.0 years, respectively). These and other similar estimates here underscore the *large* potential for further major extensions of life expectancy for middle-aged adults (and the whole population) through improved life-styles leading to shift to more favorable population mean levels of these and other important risk factors. They also emphasize the importance, in efforts to estimate potential for greater life expectancy as a consequence of improved life-styles, to consider the joint potential contribution of the key components. Thus, improved nutrition, along the lines recommended in major recent reports on the prevention of prevalent CVD and other chronic diseases, has the capacity to shift distributions of multiple risk factors to more favorable levels for the whole population, including distributions of serum lipids, systolic and diastolic blood pressure, glycemia and diabetes, serum uric acid and weight [16, 25, 27–29, 38, 45, 46, 48–60, 66–70]. Estimates of potential for enhanced longevity that focus with tunnel vision on a single trait, e.g. serum cholesterol, and assume an influence of improved nutrition on this trait and this trait only, are based on a narrow and questionable assumption [71–73]. Moreover, they ignore the potential independent contribution of improved nutrition to life expectancy, over and above its influence on serum cholesterol, blood pressure and glycemia. As the WE data indicate, and as the experience of centuries confirms [74–76], nutrition is a powerful and decisive determinant of population life expectancy, operating in many ways. For this reason alone, the cited type of single factor estimate of ability to extend life expectancy is fatally flawed and needs to be rejected as an unscientific impediment to progress in medical care and public health.

The quantitative data on the Chicago cohorts, as well as those for the MRFIT men, also underscore the conclusion that – due to prevalent life-styles in the USA and most other industrialized countries in the middle of the 20th century – the population in general is at high risk. Observed usual mean values of the population for these etiologically significant risk factors cannot in any scientific sense be designated 'normal' values. Correspondingly, only a strategy that includes a population-wide approach, aimed at shifting mean risk factor levels of the whole population downward, can encompass all that needs to be done to bring epidemic coronary disease under control.

The quantitative data presented here also indicate the importance of intensive special efforts to reduce risk of persons with very high levels of the major risk factors. Thus, for those with such all too common traits as habitual dietary cholesterol intake of 400 mg/1,000 kcal, serum cholesterol of 280 mg/dl, systolic pressure of 162 mm Hg, consumption of 20 cigarettes/day, risks of mortality from CHD and all causes, and of shortened life expectancy, are considerably greater even than for those with overall population average levels of these traits, i.e. people who are – as noted above – at high (not optimal) risk. For example, for WE men with the first three of these traits manifest together at baseline, risk of CHD death during the ensuing 24 years was greater than for the whole WE cohort by 3.6 times and risk of all-cause death by 2.3 times, with estimated life expectancy shorter by about 8 years. And when risks for those with such very high levels are compared to risks for those with favorable levels, the challenge to medical care and public health to improve prognosis for very high risk people is even more starkly demonstrated. Thus, for the same WE men with very high levels of the first three traits, risks of death from CHD and all causes in 24 years were 8.8 and 3.9 times greater than for men with favorable values, and estimated life expectancy was about 14 years less. For men with these three traits very high (as stipulated) and also smoking 20 cigarettes/day, risks of CHD and all-cause death compared to men with favorable values were 13.9 and 6.9 times greater, and estimated life expectancy was about 20 years less.

From these several data sets then, the conclusion is compelling that control of the coronary epidemic requires an effectively implemented long-term *2-pronged strategy* national in scope, combining in interdigitating and complementary ways measures to reduce risk for the whole population plus special measures for very high risk people. Each prong of this strategy needs to have both community-wide aspects (community including neighborhood, schools, work sites) and aspects related to medical services for individuals and families. For the effort to be fully successful, a key component must be encouraging favorable life-styles from early childhood on, to achieve the *primary prevention of the unfavorable life-styles and life-style-related risk factors* causing epidemic CHD. This is the ultimate solution, and the only ultimate solution.

In this regard, it is relevant to emphasize the extent to which the CHD epidemic is attributable to the life-styles and life-style-related risk factors considered here, i.e. 'rich' diet, diet-related elevations of serum cholesterol and blood pressure, and cigarette smoking. For the two cohorts of men aged 25–39 at baseline (PG and CHA cohorts), RR of CHD death in 25 and 15 years were 0.274 and 0.305 – i.e. 73 and 70% lower – for men with favorable levels of serum cholesterol, systolic pressure and cigarette use, compared to the whole cohorts. The same comparison for three cohorts age 40–59 at baseline shows RR of 0.600 (PG men), 0.410 (CHA men) and 0.417 (CHA women) – i.e. risks lower by 40, 59 and 58% – for those with favorable baseline levels of the three major risk factors compared to the whole cohorts. The same comparison for two cohorts age 60–74 at baseline shows RR of 0.486 (CHA men) and 0.369 (CHA women) – i.e. risks lower by 51 and 63% compared to the whole cohort. Thus, in six of these seven comparisons, an absolute majority of CHD mortality is attributable to unfavorable population mean levels of serum cholesterol, blood pressure and cigarette use. Undoubtedly, each of these estimates is an underestimate, since it is based on only a one-time measurement of these traits, and no consideration is given to other variables such as diabetes mellitus (prevention and control of obesity lowers risk of non-insulin-dependent diabetes), physical inactivity, dietary lipid, each an apparent independent contributor to CHD risk. Thus, with four factors measured twice, a year apart, i.e. dietary cholesterol plus the foregoing three (WE men aged 40–55 at baseline), favorable levels were associated with a 25-year RR of CHD = 0.311 – i.e. lower by 69% – compared to the whole cohort; RR of all-cause mortality was 0.434 – i.e. lower by

57%, and life expectancy was estimated to be longer by 8.6 years.

Given these findings from long-term follow-up of the several Chicago cohorts, and similar results from many other prospective studies, it is puzzling still to hear and read assertions by a few colleagues that the major risk factors account for only a minority of CHD events. These assertions are contrary to fact. Their invalidity is made especially evident by the massive data on the 361,662 men screened for recruitment for MRFIT. For the 3.1% of this cohort at low risk, i.e. with serum cholesterol <182 mg/dl, blood pressure <120/<80 mm Hg, not smoking cigarettes, free of a history of treated diabetes and hospitalization for a heart attack, 12-year age-standardized CHD death rate was only 10% that of the rest of the cohort. Given the precision of this finding, based on extraordinary numbers and standardized measurement (albeit once only, hence still an underestimate due to regression dilution bias), it is reasonable to conclude that about 90% of CHD deaths of middle-aged men in an industrialized country like the USA are attributable to the cited life-styles and life-style-dependent risk factors prevalent throughout the population, hence are preventable, first and foremost through improvements in life-styles. It is further reasonable to conclude – and again the Chicago and MRFIT data support the inference – that with shift toward more favorable risk factor patterns, the population can anticipate sizable increases in life expectancy (more on that below).

Space limitations here preclude a review of the findings from randomized controlled trials – multifactorial and unifactorial – to assess effects on morbidity and mortality of intervention to reduce major risk factors. Suffice it to note that, as reviewed elsewhere [76–79], their results are generally concordant with data from prospective population studies, including those presented here, i.e. they confirm that the potential for prevention is realizable, even when intervention is begun belatedly in middle-aged or older people, after decades of exposure to unfavorable life-styles and life-style-related risk factors leading to severe atherosclerosis. Considered in their totality, the data from trials show that significant reduction in rates of CHD and CVD, as well as lower rates of all-cause mortality, can be achieved even with intervention starting only in middle-aged or older people, including – as is the case in most reported trials – those with very high risk factor levels or with a history of clinical CHD. In recent years, this conclusion – warranted previously by earlier data from trials [76–83] – has been reinforced by favorable findings in regard to prevention of CHD, CVD and total mortality with further long-term follow-up of cohorts from multifactorial trials in Norway and the USA [84–86]; seemingly unfavorable results in a Finnish trial, using multiple drugs without a standardized protocol, do not negate this conclusion [87]. This conclusion is also reinforced by reports of three additional trials confirming efficacy of diuretic-based antihypertensive treatment in preventing both cerebrovascular and cardiac events [88–90]; and by data from 15-year follow-up of participants in the Coronary Drug Project [91] and from many trials of serum cholesterol reduction by diet, diet plus drug treatment or diet plus partial ileal bypass, with serial coronary angiographic findings as end point [92–99]. In three of these latter trials, efficacy of lipid-lowering intervention has been shown not only for slowing and preventing progression of coronary atherosclerosis, and also for inducing regression of plaques, but also for decreasing incidence of clinical CHD events [94–96, 99]. In one of these, angiographically demonstrated favorable effects over the first 3 years of intervention were shown to relate significantly to lower incidence of nonfatal myocardial infarction plus CHD death during the next several years [96]. In sum, the positive findings from trials – all the more impressive given the difficulties and limitations of these studies, as noted briefly and partially above and as reviewed years ago [100] – are concordant with those from population observational studies (as well as those from animal experimentation). An extensive body of research findings of many types shows the potential for achieving prevention of coronary and CVD disease, and thereby extending longevity by control of the key unfavorable life-styles and life-style-related risk factors producing the CHD-CVD epidemic.

In the final analysis, the validity and implications of the findings presented here, and the inferences drawn from these and related data, are dependent on results from their application population wide. In this regard, the rich experience in the USA during the last 3 decades is relevant and merits brief summation here. In the USA, the effort to prevent CHD began in the late 1950s and early 1960s [51, 52, 101–104], at a time when CHD mortality rates were still rising and were at levels second highest in the world. In the mid- and late 1960s, these rates plateaued and then began to decline. They have been falling ever since, in every sector of the country, for both middle-aged and older people, men and women, black and white [105, 106]. Overall the decrease in mortality rate exceeds 50% for CHD and is even greater for stroke. During the most recent decade of record, from 1979 to 1989, age-adjusted death rate from CHD fell 30% and from stroke 33% [106]. For the age group 55–64, these declines were 33 and 28%; for the age group 65–74, 31 and 36%.

Data from local population studies indicate that the decline in CHD death reflects primarily a fall in incidence of 'hard' CHD, i.e. nonfatal myocardial infarction and fatal CHD, including sudden coronary death [107, 108]. The decreases in CHD and stroke death rates have led not only to a substantial fall in mortality from all CVD diseases; they have been the key factor in the long-term fall in all-cause mortality rates (e.g. age-adjusted death rates from all causes down 28% from 1967 to 1989) [45]. As a consequence, in the 1970s and 1980s life expectancy increased sizably for Americans, overall by 4.5 years from 1970 to 1989 for all newborns (4.7 years for white males, 3.6 years for white females, 4.8 years for black males, 5.2 years for black females). This achievement is particularly noteworthy given the much smaller increase during the preceding 2 decades (1950–1970), only 2.6 years for all newborns, due especially to the limited advance of only 1.5 years for white males. Correspondingly, for persons who attained middle age, after a period of virtually no increase in life expectancy for white and black men, and even a decline, life expectancy has been rising during the last 2 decades. Thus, for men aged 45 (both black and white) life expectancy increased by 2 years from 1970 to 1980; similar gains were also recorded for women (both black and white) [45]. Clearly, then, middle-aged American adults can experience substantial increases in life expectancy with control of the CHD-CVD epidemic, as predicted by the data presented here from prospective research on the impact of etiologically significant major risk factors in US population samples.

During these last 20–25 years, significant changes have occurred in food consumption patterns of Americans, of an extent and scope detectable in routine US national statistics, e.g. annual data on food availability and data from periodic surveys of representative population samples [76, 78, 100, 103, 104, 109]. These data and others from special research studies indicate that on average intake of dietary cholesterol and saturated fat by the

US population has gone down, while intake of polyunsaturated fat has increased. The changes have been substantial, even though not sufficient to achieve recommended first goals for the population (e.g. no more than 300 mg/day or 100 mg/1,000 kcal of dietary cholesterol, no more than 10% of total calories from saturated fat, up to 10% from polyunsaturated fat). Correspondingly, multiple data sets show that for American adults, population mean levels of serum cholesterol and prevalence rates of hypercholesterolemia have fallen, even though population mean BMI and prevalence rates of obesity have not declined [110, 111]. For middle-aged and older adults, data from successive health surveys of random samples of the US population also show lower mean blood pressure levels and lower rates of hypertension [112], probably attributable mostly to antihypertensive drug treatment, since there is only limited evidence indicating substantial favorable trends in nutritional and other life-style patterns (e.g. salt or alcohol intake, body mass) related to blood pressure. Further, sustained large decreases have occurred in prevalence of cigarette smoking [76, 78, 103, 104, 113, 114]. There is also evidence indicative of increases in leisure time physical activity [104]. It is a reasonable inference that these multiple favorable trends in life-style and life-style-related major risk factors have contributed importantly to the substantial declines in coronary, stroke, cardiovascular and all-cause mortality, and to the increases in longevity of the US population from the late 1960s to the late 1980s.

There is one special feature of these trends that has considerable significance, both in relation to the scientific validity of the conclusions presented here and in regard to the challenge lying ahead in the USA to improve life expectancy with better health for the whole population: adoption of favorable life-styles – eating and exercise habits, avoidance of tobacco – has been considerably greater among more educated-affluent than less educated-affluent Americans. A corresponding trend pattern has been recorded in regard to serum cholesterol and body mass. An inference flows from this, as hypothesis: if trends of life-style and life-style-related risk factors have contributed importantly to the decreases in CHD-CVD mortality rates in the USA, then the declines in these rates should be greater for the more than for the less educated-affluent Americans. Five reports show that this is indeed the consistent finding [107, 115–118]. Their data lend further support to the *main conclusions of this paper.*

(1) The major life-styles and life-style-related risk factors delineated here are of massive and decisive importance in the etiology of the CHD-CVD epidemic.

(2) Effective sustained further implementation of a combined two-pronged strategy – population wide and high risk – to advance the primary prevention and control of these major exposures has a vast potential to maintain and extend the favorable trends of CHD-CVD morbidity-mortality in the USA, break the back of the epidemic, and achieve further substantial increases in longevity with better health.

The flank of the CHD-CVD epidemic has been turned in the USA but rates of sickness, disability and death from these diseases are still high. Special problems and challenges exist in regard to tens of millions of less educated-affluent Americans. Application throughout the population of the knowledge on the key causes of this epidemic – made clear by the concordant evidence collected worldwide during the last decades – is the key to solving these problems, and achieving the health goals for the United States for the year 2000 [55].

Acknowledgements

We are pleased to express appreciation to the officers, executive leadership and labor force of the Western Electric Company, Chicago, Ill., USA; their cooperation made this study possible. We are also gratified to acknowledge the role of Oglesby Paul, MD, and Mark Lepper, MD, in the initiation of the WE study and in its leadership for many years, of Anne MacMillan Shryock, BS, in the collection of the nutrition data on the WE men, and of the many physicians who participated in the annual examinations of the cohort: Maurice Albala, Harry Bliss, Herschel Browns, Marvin Colbert, Henry De Young, Peter Economou, Sanford Franzblau, Robert Felix, John Graettinger, Buford Hall, Wallace Kirkland, Joseph Muenster, Hyman Mackler, Adrian Ostfeld, Robert Parsons, Charles Perlia, William Phelan, Norman Roberg, Marvin Rosenberg, George Saxton, Armin Schick, John Sharp, Jay Silverman, Donald Tarun and Walter Wood. We are also pleased to thank Daniel Garside and Lois Steinfeldt for assistance with WE data files, and Carol Maliza for supervising the 20- and 25-year follow-up surveys. Acknowledgement is also gratefully extended to Ronald Prineas, MD, PhD, and members of the ECG Coding Laboratory, Division of Epidemiology, School of Public Health, University of Minnesota, Minneapolis, Minn., USA, for the ECG coding and quality control.

It is also a pleasure to express appreciation to the officers, executive leadership and Medical Department staff of the Peoples Gas Company in Chicago, including the successive Medical Directors Howard A. Lindberg, MD, John F. Marquardt, MD, and E. Alannah Ruder, MD; their cooperation and support over the years since the late 1950s made these two studies possible. Thanks are due to the research staff of the PG study, Wilda Miller, Elizabeth Stevens, Wanda Drake, Celene Epstein and Tom Tokich.

The work of the Chicago Heart Association Detection Project in Industry was accomplished thanks to the invaluable cooperation of many Chicago area companies and organizations, their officers, staffs and employees whose volunteer efforts made this project possible. Acknowledgement is also gratefully extended to all those in the Chicago Heart Association – staff and volunteers – serving the project: Louis DeBoer and Ray Restivo, Executive Directors, and Kay Westfall and Betty Naysmith, Program Directors; the CHA staff; Thelma Black, Clarice Blanton, Joan Carothers, Pamela Desmond, Arlene Dungca, Mary Ann Foelker, Susan Forkos, Carol Fulgenzi, Harold Gram, Jean Graver, Inger Hansen, Cherry Latimer, Mary Newman, Peggy Powell, Sue Shekelle, Karen Strentz and Suzann Ward; also, the volunteer members of the Heart Disease Committee of the Chicago Heart Association and its subcommittees: Howard Adler, René Arcilla, Robert Arzoecher, Richard A. Carleton, Angelo Cottini, Edwin Duffin, Morton B. Epstein, Robert E. Fitzgerald, Philip Freedman, Burton J. Grossman, Robert R.J. Hilker, Robert S. Kassriel, Clinton L. Lindo, Gerald Masek, Richard McNamara, Robert A. Miller, Robert Mosely Jr, Milton H. Paul, Wallace Salzman, Robert Sessions, Howard H. Sky-Peck, Donald Singer, Lachichida Sinha, Grace Smedstead, Ralph Springer, J. Martin Stoker, Carl Vogel, Ira T. Whipple and Quentin D. Young.

It is also a pleasure to acknowledge the years-long contributions to the Chicago population studies of colleagues in the Department of Preventive Medicine, Northwestern University Medical School, particularly David M. Berkson, MD, Patricia Collette, MA, Dan Garside, MA, Linda Van Horn, PhD, Kiang Liu, PhD, and Karen Ruth, MS. The biochemical analyses for the CHA and PG studies were done in the special research laboratories of the Division of Adult Health and Aging, Chicago Health Department, under the direction of Howard Adler, PhD, and Morton B. Epstein, PhD; Dana King and Ika Tomashewsky gave excellent service for years as technicians.

It is also gratifying to acknowledge the indispensable contribution of the many colleagues in the 20 clinical centers across the USA where the 361,662 men were screened for MRFIT, in the laboratories where the serum cholesterol analyses were done and in the MRFIT Coordinating Center, Division of Biometry, School of Public Health, University of Minnesota, where the analyses of the MRFIT data were done. Particular thanks are extended to Marcus Kjelsberg, PhD, there, Deborah Durkin, JoAnn Shih and Deborah Wentworth.

The Chicago research has been supported by the American Heart Association and its Chicago and Illinois affiliates; the National Heart, Lung, and Blood Institute (HL 15174 and 21010); the Chicago Health Research Foundation; the Otho S. Sprague Foundation; the Research and Education Committee of the Presbyterian-St. Luke's Hospital; the Illini Foundation, and private donors. The MRFIT investigation has been a collaborative research undertaking with National Heart, Lung, and Blood Institute contract and grant support (HL 28715).

References

1 Paul O, Lepper MH, Phelan WH, Dupertuis GW, MacMillan A, McKean H: A longitudinal study of coronary heart disease. Circulation 1963;28:20–31.

2 Ostfeld AM, Paul O: The inheritance of hypertension. Lancet 1963; 1:575–579.

3 Ostfeld AM, Shekelle RB: Psychological variables and blood pressure; in Stamler J, Stamler R, Pullman TN (eds): The Epidemiology of Hypertension. New York, Grune & Stratton, 1967, pp 321–331.

4 Paul O: The natural history of hypertension; in Stamler J, Stamler R, Pullman TN (eds): The Epidemiology of Hypertension. New York, Grune & Stratton, 1967 pp 365–374.

5 Stamler J, Berkson DM, Dyer A, Lepper MH, Lindberg HA, Paul O, McKean H, Rhomberg P, Schoenberger JA, Shekelle RB, Stamler R: Relationship of multiple variables to blood pressure – Findings from four Chicago epidemiologic studies; in Paul O (ed): Epidemiology and Control of Hypertension. Miami, Symposia Specialists, 1975, pp 307–356.

6 Dyer AR, Stamler J, Paul O, Berkson DM, Lepper MH, McKean H, Shekelle RB, Lindberg HA, Garside D: Alcohol consumption, cardiovascular risk factors, and mortality in two Chicago epidemiologic studies. Circulation 1977;56:1067–1074.

7 Pooling Project Research Group: Relationship of blood pressure, serum cholesterol, smoking habit, relative weight and ECG abnormalities to incidence of major coronary events: Final report of the Pooling Project. J Chronic Dis 1978;31: 201–306.

8 Stamler R, Stamler J, Lindberg HA, Marquardt J, Berkson DM, Paul O, Lepper M, Dyer AR, Stevens E: Asymptomatic hyperglycemia and coronary heart disease in middle-aged men in two employed populations in Chicago. J Chronic Dis 1979;32:805–815.

9 Dyer AR, Persky V, Stamler J, Paul O, Shekelle RB, Berkson DM, Lepper M, Schoenberger JA, Lindberg HA: Heart rate as a prognostic factor for coronary heart disease and mortality: Findings in three Chicago epidemiologic studies. Am J Epidemiol 1980;112:736–749.

10 Shekelle RB, Shryock AM, Paul O, Lepper M, Stamler J, Liu S, Raynor WJ Jr: Diet, serum cholesterol, and death from coronary heart disease: The Western Electric study: N Engl J Med 1981;304:65–70.

11 Shekelle RB, Lepper M, Liu S, Maliza C, Raynor WJ Jr, Rossof AH, Paul O, Shryock AM, Stamler J: Dietary vitamin A and risk of cancer in the Western Electric study. Lancet 1981;ii:1186–1190.

12 Shekelle RB, Paul O, Shryock AM, Stamler J: Dietary lipids, serum cholesterol concentration, and risk of death from coronary heart disease; in Noseda G, Fragiacomo C, Fumagalli R, Paoletti R (eds): Lipoproteins and Coronary Atherosclerosis. Amsterdam, Elsevier Biomedical Press, 1982, pp 87–94.

13 Shekelle RB, Stamler J, Paul O, Shryock AM, Liu S, Lepper M: Dietary lipids and serum cholesterol level: Change in diet confounds the cross-sectional association. Am J Epidemiol 1982;115:506–514.

14 Cedres BL, Liu K, Stamler J, Dyer AR, Stamler R, Berkson DM, Paul O, Lepper M, Lindberg HA, Marquardt J, Stevens E, Schoenberger JA, Shekelle RB, Collette P, Garside D: Independent contribution of electrocardiographic abnormalities to risk of death from coronary heart disease, cardiovascular diseases and all causes. Findings of three Chicago epidemiologic studies. Circulation 1982;65:146–153.

15 Liu K, Cedres LB, Stamler J, Dyer A, Stamler R, Nanas S, Berkson DM, Paul O, Lepper M, Lindberg HA, Marquardt J, Stevens E, Schoenberger JA, Shekelle RB, Collette P, Shekelle S, Garside D: Relationship of education to major risk factors and death from coronary heart disease, cardiovascular diseases and all causes. Findings of three Chicago epidemiologic studies. Circulation 1982;66:1308–1314.

16 Stamler J, Shekelle R: Dietary cholesterol and human coronary heart disease. The epidemiologic evidence. Arch Pathol Lab Med 1988; 112:1032–1040.

17 Shekelle RB, Stamler J: Dietary cholesterol and ischaemic heart disease. Lancet 1989;i:1177–1179.

18 Abell LL, Levy BB, Brodie BB, Kendall FE: A simplified method for estimation of total cholesterol in serum and demonstration of its specificity. J Biol Chem 1952;195:357–366.

19 Burke BS: The dietary history as a tool in research. J Am Diet Assoc 1947;23:1041–1046.

20 Department of Nutrition: Food Composition table No. 3. Boston, Harvard School of Public Health, 1957.

21 Hayes OB, Rose G: Supplementary food composition table. J Am Diet Assoc 1958;33:26–29.

22 Hardinge MG, Crooks H: Fatty acid composition of food fats. J Am Diet Assoc 1958;34:1065–1071.

23 Bowes A De P, Church CF: Food Values of Portions Commonly Used, ed 8. Philadelphia, Lippincott, 1956.

24 Fowler SF, West BB: Food for Fifty, ed 3. New York, Wiley, 1950.

25 Stamler J: Cardiovascular diseases in the United States. Am J Cardiol 1962;10:319–340.

26 Stamler J, Berkson DM, Lindberg HA, Miller WA, Stamler R, Collette P: Socioeconomic factors in the epidemiology of hypertensive disease; in Stamler J, Stamler R, Pullman TN (eds): The Epidemiology of Hypertension. New York, Grune & Stratton, 1967, pp 289–320.

27 Stamler J: Lectures on Preventive Cardiology. New York, Grune & Stratton, 1967.

28 Stamler J, Berkson DM, Lindberg HA: Risk factors: Their role in the etiology and pathogenesis of the atherosclerotic diseases; in Wissler RW, Geer JC (eds): The Pathogenesis of Atherosclerosis. Baltimore, Williams & Wilkins, 1972, pp 41–119.

29 Stamler J: Primary prevention of sudden coronary death. Circulation 1975;52:258–279.

30 Stamler R, Stamler J, Schoenberger JA, Shekelle RB, Collette P, Shekelle S, Dyer A, Garside D, Wannamaker J: Relationship of glucose tolerance to prevalence of ECG abnormalities and to 5-year mortality from cardiovascular disease: Findings of the Chicago Heart Association Detection Project in Industry. J Chronic Dis 1979;32:817–828.

31 Pan WH, Cedres LB, Liu K, Dyer A, Schoenberger JA, Shekelle RB, Stamler R, Smith D, Collette P, Stamler J: Relationship of clinical diabetes and asymptomatic hyperglycemia to risk of coronary heart disease mortality in men and women. Am J Epidemiol 1986;123:504–516.

32 Liao Y, Liu K, Dyer A, Schoenberger JA, Shekelle RB, Collette P, Stamler J: Sex differential in the relationship of electrocardiographic ST-T abnormalities to risk of coronary death: 11.5 year follow-up findings of the Chicago Heart Association Detection Project in Industry. Circulation 1987;75:347–352.

33 Levine W, Dyer AR, Shekelle RB, Schoenberger JA, Stamler J: Serum uric acid and 11.5-year mortality of middle-aged women: Findings of the Chicago Heart Association Detection Project in Industry. J Clin Epidemiol 1989;42:257–267.

34 Levine JB, Zak B: Automated determination of serum total cholesterol. Clin Chim Acta 1964;10:381–384.

35 Hawk PB, Oser BL, Summerson WH: Practical Physiological Chemistry, ed 13. New York, Blakiston, 1954, p 564.

36 Hoffman WS: A rapid photoelectric method for the determination of glucose in blood and urine. J Biol Chem 1937;120:51–55.

37 Sherwin R, Kaelber CT, Kezdi P, Kjelsberg MO, Thomas HE Jr: The Multiple Risk Factor Intervention Trial (MRFIT). II. The development of the protocol. Prev Med 1981;10:402–425.

38 Multiple Risk Factor Intervention Trial Research Group: Multiple Risk Factor Intervention Trial – Risk factor changes and mortality results. JAMA 1982;248:1465–1477.

39 Multiple Risk Factor Intervention Trial Research Group (Neaton JD, Kuller LH, Wentworth D, Borhani NO): Total and cardiovascular mortality in relation to cigarette smoking, serum cholesterol concentration, and diastolic blood pressure among black and white males followed for five years. Am Heart J 1984;108:759–769.

40 Wentworth DN, Neaton JD, Rasmussen WL: An evaluation of the Social Security Administration master beneficiary record file and the National Death Index in the ascertainment of vital status. Am J Public Health 1983;73:1270–1274.

41 Kannel WB, Neaton JD, Wentworth D, Thomas HE, Stamler J, Hulley SB, Kjelsberg MO: Overall and coronary heart disease mortality rates in relation to major risk factors in 325, 348 men screened for the Multiple Risk Factor Intervention Trial (MRFIT). Am Heart J 1986;112:825–836.

42 Stamler J, Wentworth D, Neaton JD: Is relationship between serum cholesterol and risk of premature death from coronary heart disease continuous and graded? Findings in 356,222 primary screenees of the Multiple Risk Factor Intervention Trial (MRFIT). JAMA 1986;256:2823–2828.

43 Neaton JD, Grimm RH, Cutler JA: Recruitment of participants for the Multiple Risk Factor Intervention Trial. Controlled Clin Trials 1988;8:42S–53S.

44 Cox DR: Regression models and life tables. J R Stat Soc Ser B 1972;34:187–220.

45 Stamler J, Shekelle RB: Lower dietary cholesterol and saturated fat intake, lower serum cholesterol, and expected effects on coronary risk and on longevity. Lipid Rev 1989;3:89–96.

46 Stamler J, Shekelle RB: Reduction of dietary cholesterol, saturated fat intakes and serum cholesterol: Expected effects on coronary risk and on longevity; in Lewis B, Assmann G (eds): The Social and Economic Contexts of Coronary Prevention. London, Current Medical Literature, 1990, pp 3–10.

47 National Center for Health Statistics, United States life tables: United States decennial life tables, vol I, No I, DHHS Publication No (PHS) 85-1150-1. Washington, United States Government Printing Office, 1985.

48 Stamler J, Neaton JD, Wentworth DN: Blood pressure (systolic and diastolic) and risk of fatal coronary heart disease. Hypertension 1989;13:2–12.

49 Stamler J: Blood pressure and high blood pressure: Aspects of risk. Hypertension 1991;18:I-97–I-107.

50 Stamler J: Research opportunities and directions on the blood pressure problem: Combining predictors. Am J Hypertens 1991;4:646S-660S.

51 Grundy SM, Bilheimer D, Blackburn H, Brown WV, Kwiterovich PO Jr, Mattson F, Schonfeld G, Weidman WH: Rationale of the diet-heart statement of the American Heart Association: Report of the Nutrition Committee. Circulation 1982;65:839A–854A.

52 Inter-Society Commission for Heart Disease Resources, Atherosclerosis Study Group and Epidemiology Study Group: Primary prevention of the atherosclerotic diseases. Circulation 1970;42:A55–A95.

53 Office of the Surgeon General: The Surgeon General's report on nutrition and health. Washington, United States Government Printing Office, 1988.

54 National Research Council, Committee on Diet and Health, Food and Nutrition Board, Commission on Life Sciences: Diet and Health – Implications for Reducing Chronic Disease. Washington, National Academy Press, 1989.

55 US Department of Health and Human Services: Healthy People 2000: National health promotion and disease prevention objectiv mary report). DHHS Publication No. (PHS) 91-50213. Washington, United States Government Printing Office, 1991.

56 National Cholesterol Education Program: Report of the Expert Panel on population strategies for blood cholesterol reduction. US Department of Health and Human Services, Public Health Service, National Institutes of Health Publication No 90-3046, 1990.

57 National Cholesterol Education Program: Report of the NCEP Expert Panel on detection, evaluation and treatment of high blood cholesterol in adults. Arch Intern Med 1988;148:36–69.

58 WHO Expert Committee on the Prevention of Coronary Heart Disease: Prevention of coronary heart disease. Technical Report Series No 678. Geneva, World Health Organization, 1982.

59 Assmann G, Lewis B, Mancini M (Co-Chairmen) and other members of the European Study Group: Strategy for the prevention of coronary heart disease – A policy statement of the European Atherosclerosis Society. Eur Heart J 1987;8:77–88.

60 Stamler J, Neaton JD, Wentworth D, Shih J, Dyer A, Shekelle RB, Stamler R: Life styles and life style related major risk factors: Their combined impact in producing epidemic cardiovascular disease, and the potential for prevention; in Gotto AM, Lenfant C, Paoletti R, Soma M (eds): Multiple Risk Factors in Cardiovascular Disease. Norwell, Kluwer, 1992;pp 19–25.

61 MacMahon S, Peto R, Cutler J, Collins R, Sorlie P, Neaton J, Abbott R, Godwin J, Dyer A, Stamler J: Blood pressure, stroke, and coronary heart disease. 1. Prolonged differences in blood pressure: Prospective observational studies corrected for the regression dilution bias. Lancet 1990; 335:765–774.

62 Neaton JD, Blackburn H, Jacobs D, Kuller L, Lee DJ, Sherwin R, Shih J, Stamler J, Wentworth D, for the Multiple Risk Factor Intervention Trial Research Group: Serum cholesterol level and mortality: Findings for the men screened in the Multiple Risk Factor Intervention Trial. Arch Intern Med 1992;152: 1490–1500.

63 McGee D, Reed D, Yano K, Kagon A, Tillotson J: The relationship of diet to the 10-year incidence of coronary heart disease in the Honolulu Heart Program. Am J Epidemiol 1984;119:667–676.

64 Kushi LH, Lew RA, Stare FJ, Ellison CH, El Lozy M, Bourke G, Daly L, Graham I, Hickey N, Mulcahy R, Kevaney J: Diet and 20-year mortality from coronary heart disease. N Engl J Med 1985;312:811–818.

65 Kromhout D, Bosscheiter EB, Coulander CDL: The inverse relation between fish consumption and 20-year mortality from coronary heart disease. N Engl J Med 1985;312: 1205–1209.

66 Farinaro E, Stamler J, Upton M, Mojonnier L, Hall Y, Moss D, Berkson DM: Plasma glucose levels: Long-term effect of diet in the Chicago Coronary Prevention Evaluation Program. Ann Intern Med 1977;86:147–154.

67 Stamler J, Rose G, Stamler R, Elliott P, Dyer A, Marmot M: INTERSALT study findings: Public health and medical care implications. Hypertension 1989;14:570–577.

68 Law MR, Frost CD, Wald NJ: By how much does dietary salt reduction lower blood pressure? I. Analysis of observational data among populations. BMJ 1991;302:811–815.

69 Frost CD, Law MR, Wald NJ: By how much does dietary salt reduction lower blood pressure? II. Analysis of observational data within populations. BMJ 1991;302:815–819.

70 Law MR, Frost CD, Wald NJ: By how much does dietary salt reduction lower blood pressure? III. Analysis of data from trials of salt reduction. BMJ 1991;302:819–824.

71 Taylor WC, Pass TM, Shepard DS, Komaroff AL: Cholesterol reduction and life expectancy. A model incorporating multiple risk factors. Ann Intern Med 1987;106:605–614.

72 Browner WS, Westenhouse J, Tice JA: What if Americans ate less fat? A quantitative estimate of the effect on mortality. JAMA 1991;265: 3285–3291.

73 Grover SA, Abrahamowicz M, Joseph L, Brewer C, Coupal L, Suissa S: The benefits of treating hyperlipidemia to prevent coronary heart disease: Estimating changes in life expectancy and morbidity. JAMA 1992;267:816–822.

74 McKeown T: The Role of Medicine. Princeton, Princeton University Press, 1979.

75 McKeown T: The Origins of Human Disease. Oxford, Blackwell, 1988.

76 Stamler J: Established major coronary risk factors; in Marmot M (d): Coronary Heart Disease Epidemiology: From Aetiology to Public Health. Oxford, Oxford University Press, 1992, pp 35–66.

77 Stamler J: Review of primary prevention trials of coronary heart disease. Acta Med Scand Suppl 1985; 701:100–128.

78 Stamler J: Risk factor modification trials: Implications for the elderly. Eur Heart J 1988;9(suppl D):9–53.

79 Collins R, Peto R, MacMahon S, Hebert P, Fiebach NH, Eberlein KA, Godwin J, Qizilbash N, Taylor JO, Hennekins CH: Blood pressure, stroke and coronary heart disease. 2. Short-term reductions in blood pressure: Overview of randomised drug trials in the epidemiological context. Lancet 1990;335:827–838.

80 Hjermann I, Velve-Byre K, Holme I, Leren P: Effect of diet and smoking intervention on the incidence of coronary heart disease: Report from the Oslo Study Group of a randomized trial in healthy men. Lancet 1981;ii: 1303–1310.

81 Hjermann I: A randomized primary preventive trial in coronary heart disease: The Oslo study. Prev Med 1983;12:181–184.

82 World Health Organization European Collaborative Group: European collaborative trial of multifactorial prevention of coronary heart disease: Final report on the 6-year results. Lancet 1986;i:869–872.

83 Rose G: European collaborative trial of multifactorial prevention of coronary heart disease: Final report on the 6 year results (letter): Lancet 1987;i:685.

84 Holme I, Hjermann I, Helgeland A, Leren P: The Oslo Study: Diet and anti-smoking advice – Additional results from a five-year primary preventive trial in middle-age men. Prev Med 1985;14:279–292.

85 Multiple Risk Factor Intervention Trial Research Group: Mortality rates after 10.5 years for participants in the Multiple Risk Factor Intervention Trial – Findings related to a priori hypotheses of the trial. JAMA 1990;263:1795–1801.

86 Multiple Risk Factor Intervention Trial Research Group: Mortality after 10 1/2 years for hypertensive participants in the Multiple Risk Factor Intervention Trial. Circulation 1990;82:1616–1628.
87 Strandberg TE, Salomaa VV, Naukkarinen VA, Vanhanen HT, Sarna SJ, Meittinen TA: Long-term mortality after 5-year multifactorial primary prevention of cardiovascular diseases in middle-aged men. JAMA 1991;266:1225–1229.
88 SHEP Cooperative Research Group: Prevention of stroke by antihypertensive drug treatment in older persons with isolated systolic hypertension. Final results of the Systolic Hypertension in the Elderly Program (SHEP). JAMA 1991;265:3255–3264.
89 Dahlof B, Lindholm LH, Hansson L, Schersten B, Ekbom T, Wester PO: Morbidity and mortality in the Swedish Trial in Old Patients with Hypertension (STOP-Hypertension). Lancet 1991;338:1281–1300.
90 Medical Research Council Working Party: Medical Research Council trial of treatment of hypertension in older adults: Principal results. BMJ 1992;304:405–412.
91 Canner PL, Berge KG, Wenger NK, Stamler J, Friedman L, Prineas RJ, Friedewald W: Fifteen years mortality in Coronary Drug Project patients: Long-term benefit with niacin. J Am Coll Cardiol 1986;8:1245–1255.
92 Arntzenius AC, Kromhout D, Barth JD, Reiber JH, Bruschke AV, Buis B, Vangent CM, Kempen-Voogd N, Strikwerda S, Van Der Velde EA: Diet, lipoproteins, and the progression of coronary atherosclerosis: The Leiden Intervention Trial. N Engl J Med 1985;312:805–811.
93 Blankenhorn DH, Nessim SA, Johnson RL, Sanmarco ME, Azen SP, Cashin-Hemphill L: Beneficial effects of combined colestipol-niacin therapy on coronary atherosclerosis and coronary venous bypass grafts. JAMA 1987;257:3233–3240.
94 Brown G, Albers JJ, Fisher LD, Schaefer SM, Lin JT, Kaplan C, Zhao XQ, Bisson BD, Fitzpatrick VF, Dodge HT: Regression of coronary artery disease as a result of intensive lipid-lowering therapy in men with high levels of apolipoprotein B. N Engl J Med 1990;323:1289–1298.
95 Buchwald H, Varco RL, Matts JP, Long JM, Fitch LL, Campbell GS, Pearce MB, Yellin AE, Edmiston WA, Smink RD Jr, Sawin HS Jr, Campos CT, Hansen BJ, Tuna N, Kranegis JN, Sanmarco ME, Amplatz K, Castaneda-Zuniga WR, Hunter DW, Bissett JK, Weber FJ, Stevenson JW, Leon AS, Chalmers TC and the POSCH Group: Effect of partial ileal bypass surgery on mortality and morbidity from coronary heart disease in patients with hypercholesterolemia: Report of the Program on the Surgical Control of the Hyperlipidemias (POSCH). N Engl J Med 1990;323:946–955.
96 Buchwald H, Matts JP, Fitch LL, Campos CT, Sanmarco ME, Amplatz K, Castaneda-Zuniga WR, Hunter DW, Pearce MB, Bissett JK, Edmiston WA, Sawin HS Jr, Weber FJ, Varco RL, Campbell GS, Yellin AE, Smink RD Jr, Long JM, Hansen BJ, Chalmers TC, Meier P, Stamler J and the POSCH Group. Changes in sequential coronary arteriograms and subsequent clinical events: Report of the program on the Surgical Control of Hyperlipidemias (POSCH). JAMA 1992;268:1429–1433.
97 Ornish D, Brown SE, Scherwitz LW, Billings JH, Armstrong WT, Ports TA, McLanahan SM, Kirkeeide RL, Brand RJ, Gould KL: Can lifestyle changes reverse coronary heart disease? The Lifestyle Heart Trial. Lancet 1990;336:129–133.
98 Kane JP, Malloy MJ, Ports TA, Phillips NR, Diehl JC, Havel RJ: Regression of coronary atherosclerosis during treatment of familial hypercholesterolemia with combined drug regimens. JAMA 1990;264:3007–3012.
99 Watts GF, Lewis B, Brunt JNH, Lawis ES, Coltart DJ, Smith LDR, Mann JI, Swan AV: Effects on coronary artery disease of lipid-lowering diet, or diet plus cholestyramine, in the St. Thomas' Atherosclerosis Regression Study (STARS). Lancet 1992;339:563–569.
100 Stamler J: Population studies; in Levy RI, Rifkind BM, Dennis BH, Ernst ND (eds): Nutrition, Lipids, and Coronary Heart Disease – A Global View. New York, Raven Press, 1979, pp 25–88.
101 White PD, Sprague HB, Stamler J, Stare FJ, Wright IS, Katz LN, Levine SL, Page IH: A statement on arteriosclerosis, main cause of 'heart attacks' and 'strokes'. New York, National Health Education Council, 1959.
102 Central Committee for Medical and Community Programs of the American Heart Assocation. Dietary fat and its relation to heart attacks and strokes. Circulation 1961;23:133–136.
103 Stamler J: Primary prevention of coronary heart disease: The last 20 years. Am J Cardiol 1981;47:722–735.
104 Working Group on Arteriosclerosis of the NHLBI: Prevention. Report of the Working Group on Arteriosclerosis of the National Heart, Lung, and Blood Institute, vol 2. NIH Publication No 81-2035. Bethesda, US Department of Health and Human Services, Public Health Service, 1981, pp 261–422.
105 Working Group on Arteriosclerosis of the NHLBI: Population based research. Report of the Working Group on Arteriosclerosis of the National Heart, Lung, and Blood Institute, vol 2. NIH Publication No. 81-2035. Bethesda, US Department of Health and Human Services, Public Health Service, 1981, pp 157–260.
106 National Center for Health Statistics: Advance report of final mortality statistics, 1989. Monthly Vital Statistics Report, vol 40, No 8, suppl 2. Hyattsville, Public Health Service, 1992.

107 Pell S, Fayerweather WE: Trends in the incidence of myocardial infarction and in associated mortality and morbidity in a large employed population. N Engl J Med 1985;312:1005–1012.
108 Burke GL, Sprafka JM, Folsom AR, Luepker RB, Norsted SW, Blackburn H: Trends in CHD mortality, morbidity and risk factor levels from 1960 to 1986: The Minnesota Heart Survey. Int J Epidemiol 1989;18:S73–S81.
109 Putnam JJ: Food consumption, 1970–1990. Food Rev 1991;14:2–12.
110 National Center for Health Statistics and the National Heart, Lung, and Blood Institute Collaborative Lipid Group: Trends in serum cholesterol levels among US adults aged 20 to 74 years: Data from the National Health and Nutrition Examination Surveys. 1960–1980. JAMA 1987;257:937–942.
111 Harlan WR, Landis JR, Flegal KM, Davis CS, Miller ME: Secular trends in body mass in the United States, 1960–1980. Am J Epidemiol 1988;128:1065–1074.
112 National Center for Health Statistics (Drizd T, Dannenberg AL, Engel A): Blood pressure levels in persons 18–74 years of age in 1976–80, and trends in blood pressure from 1960 to 1980 in the United States. Vital and Health Statistics, Ser 11, No 234. DHHS Publication No (PHS) 86-1684. Washington, US Government Printing Office, 1986, p 34.
113 Stamler J: Epidemiology, established risk factors, and the primary prevention of coronary heart disease; in Parmley WW, Chatterjee K (eds): Cardiology, vol 2. Philadelphia, Lippincott, 1987, chap 1, pp 1–41.
114 US Bureau of the Census: Statistical Abstract of the United States. Washington, US Government Printing Office, 1990, p 123.
115 Metropolitan Life Insurance Company: Recent trends in mortality from cardiovascular diseases. Stat Bull Metrop Life Insur Company 1979;60:3–8.
116 Enstrom JE: Trends in mortality among California physicians after giving up smoking: 1950–79. Br Med J 1983;286:1101–1105.
117 Rogot E, Hrubec Z: Trends in mortality from coronary heart disease and stroke among US veterans: 1954–79. J Clin Epidemiol 1989; 42:245–256.
118 Feldman JJ, Makuc DM, Kleinman JC, Cornoni-Huntley J: National trends in educational differentials in mortality. Am J Epidemiol 1989;129:919–933.

Cardiology 1993;82:223–228

Risk Factors for Coronary Heart Disease during 25 Years of Follow-Up

The Study of Men Born in 1913

Lennart Welin[a]
Henry Eriksson[a]
Bo Larsson[b]
Kurt Svärdsudd[c]
Lars Wilhelmsen[a]
Gösta Tibblin[c]

[a] Section of Preventive Medicine, Department of Medicine, Östra Hospital, University of Göteborg, Sweden;
[b] Department of Medicine I, Sahlgrenska Hospital, University of Göteborg, Sweden;
[c] Department of Family Medicine, University of Uppsala, Sweden

Key Words
Epidemiology
Risk factors
Prospective study, long-term

Abstract
We followed 845 50-year-old men initially free of coronary heart disease (CHD) for 25 years. Blood pressure, serum cholesterol and smoking habits were significant and independent risk factors for CHD during the first 15 years of follow-up and for the whole follow-up period. CHD during the last 10 years of follow-up could be predicted by blood pressure, serum cholesterol and triglycerides and also by body mass index, but not by smoking habits in univariate analyses. Only blood pressure was an independent risk factor for CHD during the last 10 years of follow-up in multivariate analysis.

With the 'aging' of research in cardiovascular epidemiology, long-term follow-up studies increase in numbers from the pioneering Minnesota Businessmen Study [1] and the Framingham Study [2], which started in the 40s, to numerous studies ongoing today. Our study started in 1963 when 855 50-year-old men were examined. We herein report on 25 years of follow-up.

Methods

Study Population
The participants were 50-year-old men born in 1913 and living in Göteborg at the time of the first examination in 1963. Men born on days divisible by three, i.e. the 3rd, 6th, 9th day of each month, for example, were invited and 855 men (88% participation rate) were examined [3]. These men have been followed up through reexaminations in 1967, 1973, 1980 and 1988. Nonparticipants and those who have died have been followed up through hospital and outpatient records, parish registers, the Central Bureau of Statis-

Lennart Welin, MD
Department of Medicine
Östra Hospital
S-416 85 Göteborg (Sweden)

0008-6312/93/
0823-0223$2.75/0

Table 1. Risk factors for CHD at baseline (1963) among 50-year-old men during 25 years of follow-up, during the first 15 years (1964–1978) and during the last 10 years (1979–1988).

Risk factors	No CHD (n = 665)	CHD last 10 years (n = 89)	p value	CHD first 15 years (n = 91)	p value	CHD 25 years (n = 180)	p value
Systolic blood pressure, mm Hg	136.4 ± 19.2	142.6 ± 21.1	0.006	146.9 ± 28.0	>0.001	144.8 ± 24.9	<0.001
Diastolic blood pressure, mm Hg	90.5 ± 12.6	94.7 ± 14.5	0.005	96.3 ± 14.4	<0.001	95.5 ± 14.4	<0.001
Serum cholesterol mmol/l	6.31 ± 1.11	6.60 ± 1.08	0.03	6.75 ± 1.03	<0.001	6.67 ± 1.06	<0.001
Serum triglycerides mmol/l	1.22 ± 0.73	1.45 ± 1.48	0.03	1.37 ± 0.55	0.08	1.41 ± 1.11	0.01
Blood glucose, mmol/l	4.62 ± 0.86	4.61 ± 0.82	0.9	4.72 ± 0.79	0.3	4.66 ± 0.81	0.5
BMI, kg/m²	24.7 ± 3.1	25.5 ± 3.7	0.03	24.7 ± 3.0	0.99	25.1 ± 3.4	0.14
Heart volume, ml/m²	390 ± 57	394 ± 64	0.4	403 ± 67	0.05	399 ± 65	0.07

Comparison are made with those without CHD during the whole follow-up period.
Means ± SD.

tics and death certificates. Two men have been lost to follow-up, both emigrants to the US.

Baseline Data

The men were examined in the morning after fasting overnight. Body weight was measured with a balance scale to the nearest 0.1 kg with the men wearing light clothes only. Height was measured to the nearest centimeter. The body mass index (BMI) was calculated as weight (in kg) divided by height (in m^2). By the same observer, blood pressure was recorded in the right arm, with the participant seated and after a 5-min interview. A mercury sphygmomanometer with a cuff size of 12 × 23 cm was used. Smoking habits were coded as follows: 1 = never smoked, 2 = exsmokers, 3 = smoking 1–14 g/day, 4 = smoking 15–24 g/day and 5 = smoking >25 g/day. One cigarette was estimated to be equivalent to 1 g of tobacco, one cheroot to 2 g, and one cigar to 5 g. Pipe smokers were classified according to their average daily tobacco consumption. Fasting blood samples were drawn for determinations of serum cholesterol [4], serum triglycerides [5] and blood glucose [6]. The heart volume was determined from a chest X-ray and calculated (ml/m^2 body surface area) according to Jonsell [7]. The registration by the local temperance board is used as a crude index of alcohol intemperance.

Table 2. Risk factors for CHD at baseline (1963) among 50-year-old men during 25 years of follow-up, during the first 15 years (1964–1978) and during the last 10 years (1979–1988).

Risk factor	No CHD (n = 665)	
	n	%
Smoking habits		
Never smoked	171	25.7
Exsmokers	143	21.5
Smoking 1–14 g/day	226	34.0
Smoking > 15 g/day	125	18.8
Smoking, points	2.49	(1.13)
Registration by the temperance board	135	20.3
Father dead from CHD ≤75 years old	71	10.7
Mother dead from CHD ≤75 years old	52	7.8

Comparisons are made with those without CHD during the whole follow-up period.

The parents' age and cause of death were taken from death certificates, which were obtained from the parish in which the father or the mother of the participant had died.

End Points

The end point, coronary heart disease (CHD) is defined as fatal or nonfatal myocardial infarction (MI) or death from CHD (when other causes of death could be ruled out). The criteria for MI have been published [8]: central chest pain, shock or syncope suggesting MI together with a typical transaminase rise and/or the appearance of pathological Q wave or localized ST changes in the ECG. Angina pectoris is not considered an end point.

Statistical Analysis

All analyses were performed with the Statistical Analysis System at the Göteborg Computer Center. Student's t test was used to analyze differences in mean values, and differences in proportions were analyzed with the χ^2 test. For multivariate analyses, the logistic regression (not stepwise) technique was used. All statistical tests were two tailed.

Results

Among the 855 50-year-old men who participated in 1963, 10 had already had an MI. Among the remaining 845 men, 180 (21%) developed CHD, 89 during the first 15 years of follow-up (1964–1978) and 91 during the last 10 years of follow-up (1979–1988).

During the first 15 years, blood pressure and serum cholesterol were strongly and significantly related to the incidence of CHD (table 1). Heart volume tended to be larger among CHD cases than among those with no CHD ($p = 0.05$). Serum triglycerides, blood glucose and BMI were no risk factors. Smoking habits (table 2) were also strongly related to the risk of CHD, while alcohol intemperance (registration by the temperance board) was of borderline significance ($p = 0.07$). Parental death from CHD at the age of 75 or

CHD last 10 years (n = 89)		p value	CHD first 15 years (n = 91)		p value	CHD 25 years (n = 180)		p value
n	%		n	%		n	%	
27	30.3		9	9.9		36	20.2	
15	16.9		8	8.8		23	12.8	
25	28.1		43	47.3		68	37.8	
22	24.7		31	34.1		53	29.4	
2.52	(1.24)	0.9	3.14	(1.03)	<0.001	2.83	1.18)	<0.001
15	16.9	0.4	26	28.6	0.07	41	22.8	0.47
11	12.4	0.6	11	12.1	0.7	22	12.2	0.55
8	9.0	0.7	10	11.0	0.3	18	10.0	0.35

Smoking habits also as points (see Methods) with SD within parentheses.

Table 3. Multiple logistic regression of risk factors for CHD during 25 years of follow-up, during the first 15 years and during the last 10 years of follow-up.

Risk factor	CHD 25 years		p value	CHD first 15 years		p value	CHD last 10 years		p value
	β	SEM		β	SEM		β	SEM	
Systolic blood pressure	0.0185	0.0040	<0.001	0.0225	0.0054	<0.001	0.0114	0.0057	0.05
Serum cholesterol	0.0061	0.0022	0.005	0.0069	0.0028	0.01	0.0044	0.0028	0.1
Smoking habits	0.3049	0.0771	<0.001	0.6516	0.1143	<0.001	–	–	–
Serum triglycerides	0.0663	0.1029	0.5	–	–	–	0.0871	0.1174	0.4
BMI	–	–	–	–	–	–	0.0418	0.0369	0.3
Heart volume	–	–	–	0.0021	0.0021	0.3	–	–	–

Only those risk factors which were significant in univariate analyses have been included.
β = Regression coefficient.

below (father or mother) could not predict CHD among the participants (table 2).

During the last 10 years of follow-up, again blood pressure and serum cholesterol were related to the incidence of CHD (table 1). Serum triglycerides and BMI were now also related to an increased risk of CHD, while blood glucose and heart volume were not (table 1). Smoking habits in 1963 did not increase the risk of CHD during the last 10 years of follow-up (table 2) nor did alcohol intemperance or parental death from CHD.

For the whole follow-up period of 25 years, blood pressure, serum cholesterol and triglycerides were significant risk factors for CHD (table 1) as well as smoking habits (table 2).

In the multivariate analyses (table 3), only factors significant in the univariate analyses (table 1, 2) were used. Since systolic and diastolic blood pressure are closely correlated to each other, only systolic blood pressure was entered into the multivariate analyses. For the first 15 years of follow-up as well as for the whole follow-up period, systolic blood pressure, serum cholesterol and smoking habits were statistically significant and independent risk factors for CHD. For the last 10 years, systolic blood pressure was a borderline significant risk factor for CHD.

Discussion

This 25-year follow-up of 50-year-old men continues to show that high blood pressure, increased serum cholesterol and smoking are major risk factors for CHD, as we have shown previously for a shorter follow-up period [9]. These risk factors were also the statistically strongest risk factors during the first 15 years of follow-up. During the last 10 years of follow-up, when baseline measurements herein presented were made more than 15 years before the end point events, high blood pressure and serum cholesterol remained as significant risk factors, but not smoking habits. Serum triglycerides and overweight (measured as BMI) were also risk factors for CHD during the last 10 years of follow-up in univariate analyses.

The high participation rate (88%) and almost no losses to follow-up during 25 years (2 men, 0.2%) make our findings reliable and generalizable to the general population of white middle-aged men (at least in Sweden).

It is somewhat surprising that overweight and high serum triglycerides change from being no risk factor during the first 15 years of follow-up to a significant risk factor during the last 10 years. The statistical significance is weak, however, and may well be a chance finding because multiple comparisons are made, and we may have a mass significance problem. There are, however, other findings from our study that make high triglycerides a probably true risk factor; we have found [10] that when we used baseline examinations at the ages of 60 and 67 years, high triglycerides turned out to be the strongest risk factor for CHD among these men born in 1913.

Other prospective studies usually confirm that similar CHD risk factors operate during long-term follow-up (20 years or more) as during shorter follow-up periods [11–15]. Blood pressure was the only risk factor in our study which was predictive of CHD both during the first 15 and during the last 10 years of follow-up and also an independent risk factor (the multivariate analyses). These findings are similar to those of the Dutch and Italian cohorts of the Seven Countries Study [12], the Finnish cohorts from that study [15] and the Framingham Study [14].

With respect to serum cholesterol, our findings are similar to the Dutch cohort of the Seven Countries Study (the Zutphen Study [13]); serum cholesterol was strongly predictive of CHD during the first 15 years of follow-up, but only weakly so during the last 10 years of follow-up. Most other long-term studies list cholesterol as a significant risk factor after more than 20 years of follow-up [11, 12, 14, 15].

Obesity tended to be more important as a risk factor during the last 10 years of follow-up in our study in univariate, but not in multivariate analyses. This is partly in agreement with findings from the Framingham Study [14], in which it was found that obesity was more important as a risk factor for CHD during long-term than during short-term follow-up.

In our study as well as in other long-term follow-up studies [11, 12, 14, 15], smoking habits remained as a significant risk factor. When the follow-up period, however, is divided into early and late parts, smoking habits (at baseline) were of little predictive importance for CHD during the later part of the follow-up period in Finland [15] as well as in our study. This is probably due to the men giving up smoking or dying from smoke-related (but noncardiac) diseases, like lung cancer and chronic bronchitis. When the men in our study have died from these diseases (without previous MI), they have been counted as non-CHD cases in the statistical analyses.

Parental death from CHD, a possible hereditary factor, was not predictive of CHD in our study. This does not exclude that hereditary factors like familial hypercholesterolemia are of importance, especially at younger ages.

A record of alcohol intemperance was a risk factor for CHD in our cohort during 9 years of follow-up [16], but not when the follow-up period was extended to 15 and 25 years.

Our study thus confirms that the traditional risk factors for CHD (blood pressure, cholesterol and smoking) can predict CHD even during very long follow-up periods.

References

1 Keys A, Taylor HL, Blackburn H, et al: Coronary heart disease among Minnesota business and professional men followed 15 years. Circulation 1963;28:381–395.

2 Dawber TR, Meadors GF, Moore FE: Epidemiological approaches to heart disease: The Framingham Study. Am J Public Health 1951;41: 279–286.

3 Tibblin G: High blood pressure in men aged 50. Acta Med Scand Suppl 1967;470:1–84.

4 Cramér K, Isaksson B: An evaluation of the Theorell method for the determination of total serum cholesterol. Scand J Clin Lab Invest 1959; 11:213–216.

5 Carlsson LA, Wadström LB: Determination of glycerides in blood serum. Clin Chim Acta 1959;4:197–199.

6 Levin K, Linde S: Determination of glucose in blood, cerebrospinal fluid and urine with a new glucose oxidase reagent. Svenska Läkartid 1962;59:3016–3026.

7 Jonsell SA: A method for the determination of the heart size by teleroentgenography (a heart volume index). Acta Radiol 1939;20:325–340.

8 Elmfeldt D, Wilhelmsen L, Tibblin G, et al: Registration of myocardial infarction in the city of Göteborg, Sweden. J Chronic Dis 1975;28: 173–186.

9 Tibblin G, Wilhelmsen L, Werkö L: Risk factors for myocardial infarction and death due to ischemic heart disease and other causes. Am J Cardiol 1975;35:514–522.

10 Welin L, Eriksson H, Larsson B, et al: Triglycerides, a major coronary risk factor in elderly men. A study of men born in 1913. Eur Heart J 1991; 12:700–704.

11 Keys A, Taylor HL, Blackburn H, Brozek J, Anderson JT, Simonson E: Mortality and coronary heart disease among men studied for 23 years. Arch Intern Med 1971;128: 201–214.

12 Menotti A, Keys A, Kromhout D, et al: Twenty-five year mortality from coronary heart disease and its prediction in five cohorts of middle-aged men in Finland, the Netherlands, and Italy. Prev Med 1990;19: 270–278.

13 Kromhout D, Bosschieter EB, Drijver M, Coulander C de L: Serum cholesterol and 25-year incidence of and mortality from myocardial infarction and cancer. The Zutphen Study. Arch Intern Med 1988;148:1051–1055.

14 Hubert HB, Feinleib M, McNamara PM, Castelli WP: Obesity as an independent risk factor for cardiovascular disease: A 26-year follow-up of participants in the Framingham Heart Study. Circulation 1983;67: 968–977.

15 Pekkanen J, Nissinen A, Puska P, Punsar S, Karvonen MJ: Risk factors and 25 year risk of coronary heart disease in a male population with a high incidence of the disease: The Finnish cohorts of the Seven Countries Study. BMJ 1989;299: 81–85.

16 Wilhelmsen L, Wedel H, Tibblin G: Multivariate analysis of risk factors for coronary heart disease. Circulation 1973;48:950–958.